JOURNAL OF GREEN ENGINEERING

Volume 2, No. 3 (April/May 2012)

Special issue on

Exploiting ICT for a Sustainable Green World

Guest Editors:

Panagiotis Demestichas, Klauss Moessner and Djamal Zeghlache

JOURNAL OF GREEN ENGINEERING

Aims and Scopes
Journal of Green Engineering will publish original, high quality, peer-reviewed research papers and review articles dealing with environmentally safe engineering including their systems. Paper submission is solicited on:

- Theoretical and numerical modeling of environmentally safe electrical engineering devices and systems.
- Simulation of performance of innovative energy supply systems including renewable energy systems, as well as energy harvesting systems.
- Modeling and optimization of human environmentally conscientiousness environment (especially related to electromagnetics and acoustics).
- Modeling and optimization of applications of engineering sciences and technology to medicine and biology.
- Advances in modeling including optimization, product modeling, fault detection and diagnostics, inverse models.
- Advances in software and systems interoperability, validation and calibration techniques. Simulation tools for sustainable environment (especially electromagnetic, and acoustic).
- Experiences on teaching environmentally safe engineering (including applications of engineering sciences and technology to medicine and biology).

All these topics may be addressed from a global scale to a microscopic scale, and for different phases during the life cycle.

JOURNAL OF GREEN ENGINEERING

Volume 2   No. 3   April/May 2012

Published, sold and distributed by:
River Publishers
P.O. Box 1657
Algade 42
9000 Aalborg
Denmark

Tel.: +45369953197
www.riverpublishers.com

Journal of Green Engineering is published four times a year.
Publication programme, 2011–2012: Volume 2 (4 issues)

ISSN 1904-4720

Editorial Foreword – Exploiting ICT for a Sustainable Green World

Panagiotis Demestichas[1], Klaus Moessner[2] and Djamal Zeghlache[3]

[1]*University of Piraeus, Greece; e-mail: pdemest@unipi.gr*
[2]*University of Surrey, U.K.; e-mail: k.moessner@surrey.ac.uk*
[3]*Télécom SudParis, France; e-mail: djamal.zeglache@it-sudparis.eu*

During the last years social challenges such as pollution, dwindling natural resources and climate change are being more intensively taken into consideration by federal and local governments, public and private industry, citizens and in a wider sense by the European Union.

This is also certified by the fact that current investments by the European Union in cleaner technologies is increasing aggressively to meet carbon-reduction targets, boost energy security, and improve the economic sustainability and health of its communities. Private capital investors seek to invest in companies with innovations that exploit the commercial potential of green technology. Private industry and public organizations are investing in research and development and producing new technologies that address the demands set by legislative policies and consumers. Researchers and technologists are developing new solutions that improve the efficiency and efficacy of existing environmental technologies and infrastructures. Consumers are demanding that the products they buy and services they use are environmentally sound. These green technology stakeholders are engaged in parallel, and not often, in collaboration on ways to best address the demands and challenges of this fast-changing green technology challenge and market.

Therefore, green technologies are continuously gaining industry, research, government and non-governmental attention. So, obviously, now is the appropriate time for taking the essential steps that will show that the involved parties are highly active in the area, along with the ICT solution they have developed.

For this issue, we received high quality contributions. After extended impartial and rigorous review process by independent reviewers, we accepted six papers. The papers selected for this issue cover the areas of achieving energy efficiency in core networks, making use of opportunistic networks and also using self-organisation in network sharing. Moreover, the economic perspective of energy efficiency is discussed, while the issue is concluded with energy efficiency in the residential and business environments. Next, we briefly provide a high-level overview of each accepted paper.

V. Foteinos et al. present the "Energy Savings with Multilayer Traffic Engineering in Future Core Networks". In this article, the way multilayer traffic engineering techniques are used in order to help increasing energy savings and minimizing energy consumption. The simultaneous optimization of routing and minimization of the number of active network elements leads to significant energy savings. An adaptive algorithm for IP/MPLS/WDM with traffic grooming networks is proposed, which aggregates traffic flows in the same LSPs and consequently in the same optical wavelengths. Additionally, this algorithm has the objective to reduce energy by switching off unused router line cards. In this way routing is optimized and the topology of IP layer is adapted to the variations of traffic in network. To achieve that, LSPs (links and optical channels) must be recomputed every time a new traffic flow enters network, using a specific cost function to compute link and channels weights according to their current availability.

In S. Mumtaz et al. the "Green ICT: Self-Organization Aided Network Sharing in LTEA" is presented. This paper provides a tutorial on a new energy reduction approach – self-organization aided network sharing, which exploits the concept of cooperation and adaptation in the cellular system level. Detail reviews from different perspectives including the academic literature, the existing standard, etc. are provided. Moreover, a simply example with simulation results is provided to demonstrate the energy efficiency of the proposed scheme.

In the next paper D. Karvounas et al. present "Achieving Energy Efficiency through the Opportunistic Exploitation of Infrastructures Comprising Cells of Various Sizes". This article presents the concept of Opportunistic Networks (ONs) as an energy efficient method to exploit wireless networks. The cases studied in this work comprise a macro base station and femtocells. The proposed solution will offload a proportion of the traffic of the base station to the femtocells, through the creation of an ON. Therefore, the base station will consume less energy since terminals will be rerouted to the femtocells. In addition, the femto-terminals will operate to lower power levels leading to battery savings. Results from simulations are provided to confirm the proposed solution.

Berl et al. present the "Survey on Energy Efficiency in Office and Residential Computing Environments". More specifically, this paper gives an overview on energy saving methods that are applied today, with a special focus on office and residential environments. Currently used methods are classified into three categories: (1) autonomous management of devices, (2) coordinated management of devices, and (3) coordinated management of services. Various implementations of these methods in office and residential environments are described and compared to each other. The comparison illustrates possible directions of future research in the area of energy efficiency.

The perspective of "Environmental and Economically Sustainable Cellular Networks" is presented by W. Guo et al. in the next paper. This paper considers the wireless cellular network for both outdoor and indoor environments. Novel and theoretical bounds are presented for energy and cost savings. Investigation results show that with careful redesign of the cellular network architecture, up to 60–70% reduction in energy consumption and 25% in OPEX can be achieved for indoor and outdoor environments. Furthermore, a detailed sensitivity analysis is also presented, which is novel and beneficial to future researchers.

Finally, S. Louca et al. present the "Ecological Alertness of Cypriot Businesses". The focus of this article is to investigate the responsiveness of Cypriot businesses in the global call for ecologically friendly initiatives along with their impact on their performance, on their business strategy and on their consumers. The findings show that companies in Cyprus are pursuing various environmental initiatives involving mainly their recycling habits and issues, certification and energy preservation.

Last but not least, the guest editors thank the authors for their contributions and the reviewers for their timely reviews and comments. Professor Demestichas would also like to thank Dr. Y. Kritikou for the help in administering the bulk of submissions and reviews. We would also like to thank the Editor-in-Chief, Professor Dina Simunic, for her constant support. We hope that the readers will enjoy this issue!

Biographies

Panagiotis Demestichas is a Professor at the University of Piraeus, Department of Digital Systems, which he joined in September 2002 as Assistant Professor. From January 2001 until August 2002 he was lecturing at the National Technical University of Athens (NTUA). From January 1997 until August 2002 he was senior research engineer in the Telecommunications Laboratory of NTUA. Until December 1996 he had acquired a Diploma and Ph.D. degree (1996) in Electrical and Computer Engineering from NTUA. He has been actively involved in a number of national and international research and development programs. His research interests include the design and performance evaluation of high-speed, wireless and wired, broadband networks, network management, software engineering, algorithms and complexity theory, and queuing theory. Currently he is the Project Coordinator of the ICT OneFIT (Opportunistic Networks and Cognitive Management Systems for Efficient Application Provision in the Future Internet) Project and he serves as the deputy leader of the Unified Management Framework workpackage in the ICT UniverSelf Project. Moreover, he was the technical manager of the "End-to-End Efficiency" (E3) project, which was on the introduction of cognitive systems for the wireless world, and was partially funded by the European Commission, under the 7th Framework Programme (FP). He is the chairman of Working Group 6 (WG6), titled "Cognitive Wireless Networks and Systems", of the Wireless World Research Forum (WWRF). He has experience in project management from the FP5 project MONASIDRE ("Management of Networks and Services in a Diversified Radio Environment"), which was on managing composite wireless networks that include mobile, WLAN and digital broadcasting networks. He has more than 150 publications in international journals and refereed conferences. He is associate editor of the IEEE Communication letters and on the Board of Editors of the *Journal of Network and Systems Management*.

Klaus Moessner is a Professor in the Centre for Communication Systems Research at the University of Surrey, UK. Klaus earned his Dipl-Ing (FH) at The University of Applied Sciences in Offenburg, Germany, an MSc from Brunel University, UK and his PhD from the University of Surrey (UK). His research interests include reconfigurability of the different system levels, including reconfiguration management, service platforms, service oriented architectures and IMS, as well as scheduling in wireless networks and adaptability of multimodal user interfaces. He is involved in investigation and teaching of mobile service platforms, service oriented architectures, mobile service delivery and service enablers. He is involved in a number of European Community supported research projects in the area of the "Internet of Things". This includes the OUTSMART project (http://www.fi-ppp.eu/projects/outsmart/) and iCore (http://www.iot-icore.eu/) where he acts as deputy technical manager. He is project coordinator of the IoT.est project (http://www.ict-iotest.eu).

Djamal Zeghlache graduated from SMU in Dallas, Texas in 1987 with a PhD in Electrical Engineering and joined the same year Cleveland State University as an Assistant Professor. In 1990 and 1991 he worked with the NASA Lewis Research Centre on mobile satellite terminals, systems and applications. In 1992 he joined the Networks and Services Department at Télécom SudParis of Institut Telecom where he currently acts as Professor and Head of the Wireless Networks and Multimedia Services Department. Professor Zeghlache is also acting Dean of Research of Tlcom SudParis. He co-authored around one hundrer publications in ranked international conferences and journals and was an editor for IEEE Transactions on Wireless. He is lead scientist for Institut Télécom in several European and national funded projects: ITEA 2 project Easi-clouds, FP7 project SAIL on Cloud Networking, and national projects Magellan on virtual clusters and cloud computing and CompatibleOne aiming at producing an open source cloudware. His interests and research activities span a broad spectrum related to fixed and wireless networks and services. The current focus is on network architectures, protocols and interfaces to ensure smooth evolution towards loosely coupled future Internet, cloud networking and cloud architectures. He is currently addressing inter-domain cooperation and federation challenges for these networks, related modelling for resource optimisation of infrastructures and platforms offered as a service to users and providers while taking into account multiple constraints, objectives and criteria such as security, protection and energy efficiency.

Energy Savings with Multilayer Traffic Engineering in Future Core Networks

Vassilis Foteinos, Kostas Tsagkaris, Pierre Peloso, Laurent Ciavaglia
and Panagiotis Demestichas

[1]*Department of Digital Systems, University of Piraeus, 18534 Piraeus, Greece;
e-mail: {vfotein, ktsagk, pdemest}@unipi.gr*
[2]*Alcatel-Lucent Bell Labs France; e-mail: {pierre.peloso,
laurent.ciavaglia}@alcatel-lucent.com*

Received 3 March 2012; Accepted: 6 April 2012

Abstract

The main advantage of multilayer automatically switched IP-over-MPLS-over-optical networks is their ability and their flexibility to adapt to the variations of the offered traffic. Multilayer traffic engineering mechanisms exploit this characteristic and perform traditional rerouting enriched with on-line IP logical topology reconfiguration. These mechanisms aim at optimizing resource usage and maximizing QoS, e.g. in terms of bandwidth, throughput and delay. Recently except from performance, energy efficiency in backbone core networks has gained much attention. In this article, we will examine how multilayer traffic engineering techniques can help to increase energy savings and minimize energy consumption. As we will present the simultaneous optimization of routing and minimization of the number of active network elements leads to significant energy savings. We will propose an adaptive algorithm for IP/MPLS/WDM with traffic grooming networks, which will aggregate traffic flows in the same LSPs and consequently in the same optical wavelengths. Additionally, this algorithm has the objective to reduce energy by switching off unused router line cards. In this way routing is optimized and the topology of IP layer is adapted to the variations of traffic in network. To achieve this, LSPs (links and optical channels) must be recomputed

Journal of Green Engineering, Vol. 2, 195–214.

every time a new traffic flow enters network, using a specific cost function to compute link and channels weights according to their current availability.

Keywords: multilayer traffic engineering, traffic grooming, energy efficiency, load balancing.

1 Introduction

Nowadays, there are advanced applications that demand instant and massive network bandwidth such as video conferencing, online high definition TV and online gaming with high quality graphics. It is clear that due to their popularity, the requested demand for bandwidth from this kind of applications will be enormous even for the core network. At the same time there has been amazing progress in the capacity of optical technologies. A wavelength may operate at up to 100 Gbps and dense WDM allows one hundred wavelengths per optical fiber in the C-Band. Therefore, the available bandwidth (over a terabit) is outsized compared to the requests. As a result, only a fraction of each wavelength might be used. Traffic grooming refers to techniques used to combine low-speed traffic streams onto wavelengths in order to use this bandwidth more efficient. Consequently, the problem is not the availability of bandwidth *per se*, but its efficient management. Hence, IP/MPLS over WDM with traffic grooming and multilayer traffic engineering mechanisms is regarded as the best solution for next generation core networks.

A multilayer IP-over-optical network combines IP/MPLS forwarding from upper layer and optical data transport and switching from lower layer. With GMPLS signaling and with optical switching technology such as optical cross-connects (OXCs) it is possible to set up and tear down paths, though the utilization of these techniques in a dynamic context is yet infrequent. In multilayer networks, IP/MPLS traffic flows rerouting can end up to a complete logical reconfiguration of IP/MPLS topology. This logical topology consists of lightpaths which could afford the capacity to adapt to the variations of traffic demands by appropriate setup and teardown of the latter.

Traffic engineering in multilayer networks aims at optimizing resource usage and bandwidth utilization. Recently, multilayer traffic engineering paid attention to another important objective namely, energy efficiency. Presently, 7–8% of the world energy consumption is absorbed by the Information and Communications Technology (ICT) infrastructures, with the Internet being responsible for about 25% of this amount. Since networks consume such a large and increasing amount of energy, "green" strategies are desirable for a more energy-efficient operation of networks. Reducing the amount of network resources needed for a certain traffic demand allows maximizing energy savings and at the same time minimizing operational expenses such as lightpath leasing fees, or even electrical power cost, which can lead to a significant portion of the operational expenditures (OPEX). Therefore one of the network operators' main objectives is to provide energy-aware network operation. Unfortunately, current underlying multilayer network infrastructures, lack effective energy management. Management solutions that would assist in managing and controlling networks and services in an energy-efficient and flexible way are of utmost importance. Taking into account the afore-mentioned major objectives, in this paper we formulate, solve and validate by simulations a general problem for energy-aware multilayer traffic-engineering in future core networks.

The rest of this paper is organized as follows. Section 2 provides related work regarding multilayer traffic engineering algorithms for optimizing network performance and saving energy. The details of our problem statement and the proposed algorithm for solving it are given in Section 3. Section 4 contains the evaluation and simulation results of our proposed algorithm. A conclusion is given in Section 5.

2 Related Work

2.1 Multilayer Traffic Engineering

Usually, in an IP/MPLS/WDM network, traffic flows have significantly lower capacity requirements than the offered bandwidth from the wavelength channels. Also they belong to different kind of services and they may require different bandwidth granularities. Unavoidably, in order to operate networks efficiently, these flows must be multiplexed onto the wavelength channels (traffic grooming). This technique can be an ideal solution for the efficient management of the available network resources [12, 16].

In this manner, the way of grooming low-speed connection requests optimally into a high-capacity lightpath is an important issue. In [1], two layer mesh networks are considered and it is examined how to set up lightpaths in the optical layer to accommodate the connection requests of the IP/MPLS layer. The objective of this paper is to maximize a utility function for an Internet service provider (ISP) under resource limitation. In [9], in order to make full use of idle resources of different network layers, two integrated optimization modes are proposed for a three-layer dynamic network. Additionally, in order to transport each service in a synchronous manner, an ant colony optimization (ACO) based routing approach is adopted in these two modes.

Another design problem in multilayer networks is the optimization of node's capacity. The goal is to optimize the capacity of LSRs and OXCs, rather than the links capacity at each network layer [6].

In [14] the authors introduce an explicit architecture for IP/MPLS-over-OTN-over-DWDM (three layer network). They present a detailed network optimization model for the operational planning of such an environment that considers OTN as a distinct layer with defined restrictions. The objective in this model is to minimize the total network cost.

The authors in [15] concentrate on making the core network efficient for transporting differentiated service traffic, adaptive to changes in traffic patterns and resilient against possible failures. To this end they introduce MPLS TE and DiffServ QoS functionality in their network model, thus ensuring maximum utilization of resources and automated guarantees of optimal quality for different types of transported traffic. In [16], to support diverse sources in an optical network, dynamic traffic grooming is needed to ensure different traffic performance objectives for different services. This may either be done by restricting bandwidth access for some of the traffic classes or by giving priority access to one type of traffic over another.

2.2 Energy Efficiency in Multilayer Networks

Undoubtedly, optimization of network energy consumption is related with optimization of network resource usage. In [11], the authors provide a comprehensive survey of the most relevant research activities for minimizing energy consumption in telecom networks, with a specific emphasis on those employing optical technologies. In optical layer there are two techniques that help in minimizing energy consumption. As we can see in [8, 10] optical bypass and traffic grooming combined with multilayer traffic engineering provide more efficient network resource usage and more energy savings from unused network elements.

In [4] the interaction of multilayer traffic engineering with hardware-based energy efficiency optimization techniques is examined. First, the authors look at scaling back power requirements through the use of better chip technology, but also decreasing idle-power requirement only, using improved chip architecture. Secondly, as multilayer traffic engineering allows for fast responses to changing traffic, they present how link switch-off during off-peak hours offers a straightforward option to reduce energy needs. These two types of power reduction techniques and their interaction with multilayer traffic engineering mechanisms are also examined in [13].

Generally, the most common way to maximize energy savings is by adapting the IP topology to the traffic that is actually carried in the network. In [5] a dynamic reconfiguration of a core IP/MPLS-over-WDM network is studied for adaption to traffic patterns changing during the course of the day. This approach tries to route the traffic preferably in the optical layer and deactivates unused IP router ports. In [7] the authors propose a mechanism, which optimizes the traffic routing and adapts the IP topology to the traffic that is actually carried in the network.

Another approach in optimizing energy consumption in networks is by optimizing node capacity since LSRs with high capacity and complex structures consume significant power. In [3], the authors present an optimization model that considers each layer's constraints (IP/MPLS over OTN over DWDM), aiming to reduce the capacity of the routing and switching nodes. The model proposed in [6] also aims to optimize the capacity of LSRs and OXCs, rather than the links capacity at each network layer.

In this paper, we will examine how multilayer traffic engineering facilitates the adaptation of network's logical topology to traffic variations in order to activate fewer network elements. We will propose an energy efficient algorithm, which uses a cost function to find the minimum cost in links and optical channels routing configuration.

3 Problem Statement and Solution

3.1 Mathematic Formulation

The proposed cost function considers the number of additional links and channels needed to support the new traffic flow. Activation of an unused link is considered more energy consuming than the activation of an unused channel. Specifically, creating a new IP link between two routers is more costly than augmenting its capacity by activating optical channels. All possible paths are assigned with a total cost depending on its links and channels available bandwidth. If one path cannot accommodate the requested traffic then its cost is assumed infinite, in order to be excluded. Otherwise, its cost consists of the number of links and channels that must be activated. Each element adds a cost equal to 1 if they must be activated or 0 if they are already activated. If a

Table 1   Description of variables used.

Variables	Description	Variables	Description
req	Requested capacity	N_l	Number of additional links
C_{pi}	Cost of path Pi	N_c	Number of additional channels
Cl	Cost of link l	P_{i_av}	Availability of path P_i
Cc	Cost of channel c	l_{av}	Availability of link l
w	Weight of channels	l	Link

link cannot support the request then its cost is infinite.

$$C_{P_i} = \begin{cases} \infty, \text{ if } P_{i\_av} < \text{req} \\ \sum_{j=\text{Ingress}}^{\text{Egress}} Cl_j + w \cdot \sum_{j=\text{Ingress}}^{\text{Egress}} Cc_j \end{cases} \quad 0 \le w \le 1,$$

$$Cl = \begin{cases} \infty, & \text{if } l_{av} < \text{req} \\ 1, & \text{if } l_{av} \ge \text{req } \& \& \ l \text{ closed} \\ 0, & \text{if } l_{av} \ge \text{req } \& \& \ l \text{ open} \end{cases} \quad Cc = N_c, \quad 0 \le Cc \le 2$$

In Table 1 we define the variables used in the proposed solution.

3.2 Analysis

Each network element (links and channels) has availability, which is the amount of free bandwidth that can offer for new requests. For a new request, every possible path is evaluated and is assigned with a cost value depending on whether it can accommodate new traffic or not and if it can how many unused network elements will be activated. In order to reduce energy consumption our algorithm selects one path that will activate the least number of links and channels, with links being more significant. Traffic flows will aggregate in the same links through LSPs and in the same channels through any port aggregation mechanism. At the end, with RSVP-TE signalling this path is established.

Generally in core networks, the basic task of resource optimization is to optimally assign the new incoming traffic one by one so that the possibility of accommodating further incoming traffic without congestion can be maximized. Toward this end, online multilayer traffic engineering approaches make sure that the traffic load is as evenly distributed as possible within the network, so random incoming traffic demand in the future can easily be satisfied. In some cases it is also possible to reroute existing flows in the network so as to reserve bandwidth for new and future incoming traffic [19].

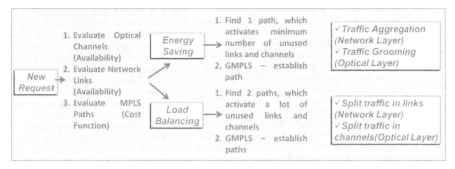

Figure 1   Energy saving – load balancing.

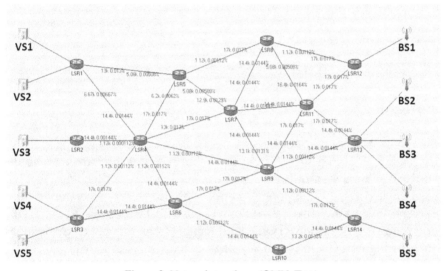

Figure 2   Network topology (OMNeT++).

In this manner, in order to evaluate the performance of our energy-efficient algorithm in realistic scenarios, we develop another approach which will have the objective of load balancing in the network. For this purpose, traffic will split in different paths and channels. Figure 1 depicts these two approaches.

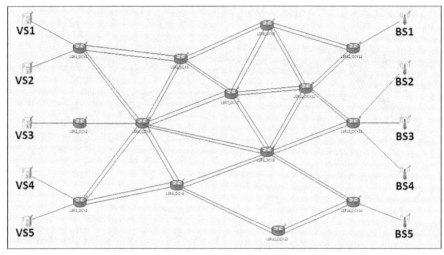

Figure 3 Network physical topology.

4 Results

4.1 Simulation Setup

Evaluation of our energy efficient algorithm has been performed using the OMNeT++ simulator. In Figure 2 we can see the network topology that has been created for this purpose. It consists of five traffic generators (Video Servers), five traffic sinks (Base Stations) and 14 LSRs-over-OCXs. Traffic flows are traversing network in the direction from sources to destinations only. In Figure 3 the physical topology of our network is depicted. Links in this network consist of two optical channels with capacity of 51.84 Mbps (base rate bandwidth) each.

4.2 Comparisons

For our evaluation we examined three scenarios. Three Base Stations (BSs) derive requests to Video Servers (VSs) and three traffic flows (indicated by the appropriate numbering) are generated respectively. In the first scenario, BSs require streams of 20 Mbps each. In the second scenario these bandwidth requirements are doubled (40 Mbps) and for the third scenario they are doubled again (80 Mbps).

In Figures 4 and 5, the logical topology configuration for the first scenario is shown for both approaches. In Figure 4 our algorithm is responsible for

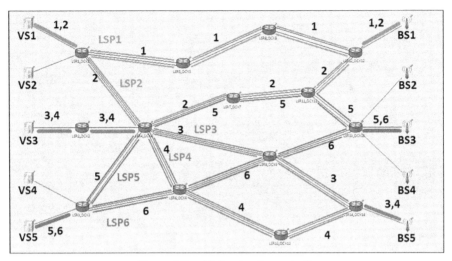

Figure 4 Load balancing (scenario 1).

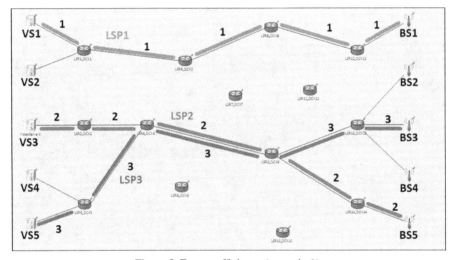

Figure 5 Energy efficiency (scenario 1).

splitting traffic among different paths and channels while in Figure 5 our algorithm aggregates traffic flows in same links and channels (traffic grooming). Unused network elements are allowed to deactivate in order to decrease energy costs. It is evident that the energy efficient routing configuration demands only a small number of links and channels especially in contrast to the load balancing approach.

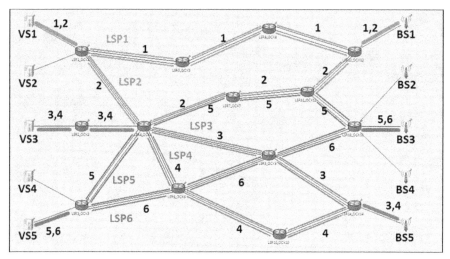

Figure 6 Load balancing (scenario 2).

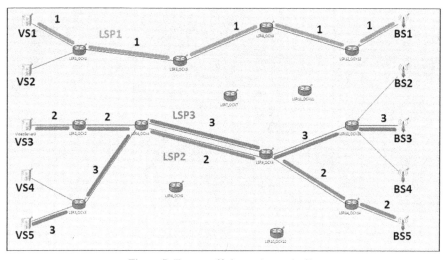

Figure 7 Energy efficiency (scenario 2).

Figures 6 and 7 depict how logical topology and routing configuration is affected by the increase in bandwidth requirements. In case of load balancing there is no need for changes due to underutilized links and channels. In case of energy efficiency despite the increased demands only one channel is required to be activated with obviously a minimum increase in energy costs.

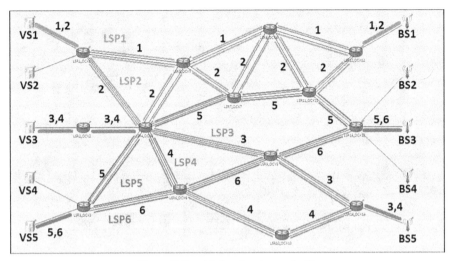

Figure 8 Load balancing (scenario 3).

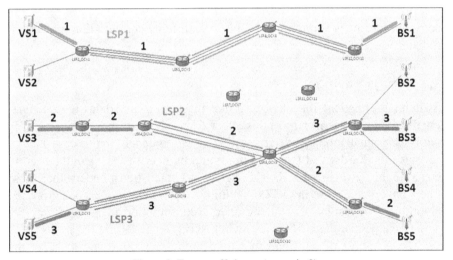

Figure 9 Energy efficiency (scenario 3).

In Figures 8 and 9 we can see how further increase in bandwidth requests affects routing configuration and therefore the logical topology. In order to avoid congestion and decrease in QoS that users experience, for some traffic flows new LSPs are set up through RSVP-TE signalling while old LSPs are torn down. Due to our energy efficient algorithm the necessary amount of

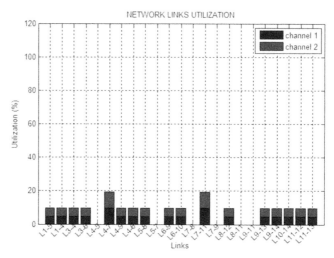

Figure 10 Links utilization – load balancing (scenario 1).

activated resources is minimized (Figure 9). Energy consumption in network remains low.

4.3 Energy Savings

Switching off optical fibers has the disadvantage of a large delay for reactivating them. In order to minimize this effect, fibers should remain activated even with a very small amount of traffic (utilizing one channel). In this case, energy savings are still derived from the minimum number of wavelengths used in these fibers. Furthermore, the number of activated transponders that these fibers are connected to in OCXs is minimized. In this manner, our proposed algorithm still results in reducing energy consumption in network and we could assume that these fibers are switched off.

The objective of our algorithm is to optimize routing and to adapt the logical topology of IP layer to the variations of traffic in network. Different routing configurations lead to different logical topologies and different level of utilization in network links and channels. In the next figures we can observe how our algorithm and especially the energy efficient approach influence the number and the utilization degree of links and channels in network. In all three scenarios most of the links remain deactivated and the activated ones mainly utilize one channel. In the third case where bandwidth requests are significantly increased the activation of channels is preferred from the

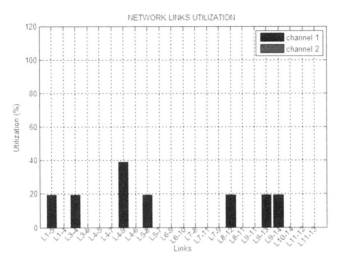

Figure 11 Links utilization – energy efficiency (scenario 1).

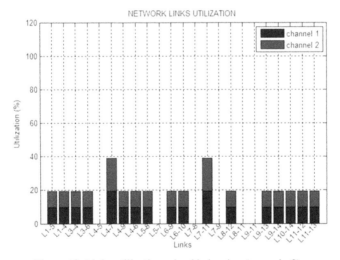

Figure 12 Links utilization – load balancing (scenario 2).

activation of links. Moreover, in these figures is demonstrated the transition between a large number of relative equally low utilized links and channels (Figures 10, 12, 14) and a smaller number of more utilized links and channels (Figures 11, 13, 15). Unavoidably, in the first case energy costs and by

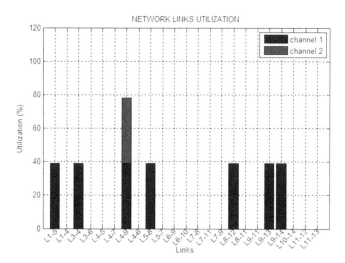

Figure 13 Links utilization – energy efficiency (scenario 2).

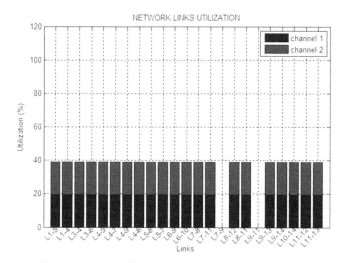

Figure 14 Links utilization – load balancing (scenario 3).

extension OPEX are increased and in the second case they are significantly lower.

In Table 2, aggregated results from these three scenarios are presented. Our energy efficient algorithm allowed the deactivation of approximately 42% of channels in comparison with the load balancing approach. The lo-

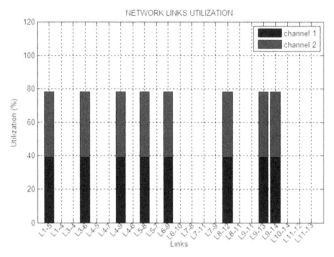

Figure 15 Links utilization – energy efficiency (scenario 3).

Table 2 Energy savings.

	Multilayer Load Balancing	Multilayer Energy Efficiency	Energy Savings
	Channels	Channels	Channels
I	40	23	42.5%
II	40	24	40%
III	44	24	45.4%

gical topology is adapted to the exact traffic that is traversing network. Traffic aggregation in conjunction with traffic grooming limited the unnecessary use of network resources.

For energy efficiency at the network layer, a key approach is to perform route selection in a way that allows large number of nodes to sleep, while still maintaining network connectivity [20]. Our mechanism can be used in this context to aggregate packets along few routes thus allowing many LSRs to sleep during periods of low load. In Table 3, we present the number of active LSRs for the two approaches we examined. With our energy efficient routing configuration, we are able to satisfy traffic demands activating minimum number of LSRs.

Aggregation of traffic in links and channels facilitate an energy efficient network operation. In the same time, it is anticipated that utilization in a

Table 3 Activated LSRs.

	Multilayer Load Balancing	Multilayer Energy Efficiency
I	14	10
II	14	10
III	14	11

Table 4 Maximum network utilization.

	Multilayer Load Balancing			Multilayer Energy Efficiency		
	Links	Channel 1	Channel 2	Links	Channel 1	Channel 2
I	19.5%	19.5%	19.5%	39%	78.1%	0%
II	39%	39%	39%	78.1%	78.1%	78.1%
III	39%	39%	39%	78.1%	78.1%	78.1%

number of links and channels will be increased. In Table 4, we present the maximum level of utilization. As it is expected, in most cases measurements show that the values are doubled.

5 Conclusions

In this paper, we examined how multilayer traffic engineering facilitates the adaptation of network's logical topology to traffic variations. Traffic aggregation, traffic grooming and the reconfiguration of LSPs can be exploited in order to activate fewer network elements. The proposed energy efficient algorithm applies these techniques and finds the minimum cost in links and channels routing configuration. This algorithm saves energy in IP/MPLS over WDM networks by switching off router line cards and uses energy more efficiently.

Reducing the amount of necessary network resources allows maximizing energy savings and at the same time minimizing operational expenses such as lightpath leasing fees for Service Providers. Additionally, lower operating costs allow an "efficient" company to gain a competitive edge over more wasteful competitors. It is worth emphasizing that these business benefits can be achieved through no capital investment.

Further research is in progress and the goal is to develop an updated cost function which will include proactive approach.

Acknowledgements

The research leading to these results has been performed within the UniverSelf project (www.univerself-project.eu) and received funding from the European Community's Seventh Framework Programme (FP7/2007-2013) under grant agreement No. 257513. Furthermore, this work has been performed in the framework of the E3 project National Participation, funded by the General Secretariat of Research and Technology (GSRT) of the Greek Ministry of Development.

References

[1] Xiaoning Zhang, Feng Shen, Li Wang, Sheng Wang, Lemin Li, and Hongbin Luo. Two-layer mesh network optimization based on inter-layer decomposition. *Photonic Network Communications*, 21(3):310–320, DOI: 10.1007/s11107-010-0301-z.

[2] Angelo Coiro, Alessandro Valenti, Francesco Matera, and Marina Settembre. Reducing power consumption in core wavelength division multiplexing optical networks. *Fiber and Integrated Optics*, 30(3):166–177, 2011.

[3] Iyad Katib and Deep Medhi. Optimizing node capacity in multilayer networks. *IEEE Communications Letters*, 15(5), May 2011.

[4] Bart Puype, Willem Vereecken, Didier Colle, Mario Pickavet, and Piet Demeester. Multilayer traffic engineering for energy efficiency. *Photon Netw. Commun.* 21:127–140, 2011, DOI 10.1007/s11107-010-0287-6.

[5] Stephan Pachnicke, Hüseyin Kagba, and Peter M. Krummrich. Load adaptive optical-bypassing for reducing core network energy consumption. PhotonischeNetze – 12. ITG-Fachtagung, Leipzig, Germany, 2011.

[6] Iyad A. Katib. IP/MPLS over OTN over DWDM multilayer networks: Optimization models, algorithms, and analyses. A dissertation in Telecommunications & Computer Networking and Computer Science & Informatics, Missouri, 2011.

[7] Angelo Coiro, Flavio Iervini, and Marco Listanti. Distributed and adaptive interface switch off for internet energy saving. In *Proceedings of 20th International Conference on Computer Communications and Networks (ICCCN)*, 31 July–4 August 2011.

[8] Francesca Vismara, Vida Grkovic, Francesco Musumeci, Massimo Tornatore, and Stefano Bregni. On the energy efficiency of IP-over-WDM networks. In *Proceedings of IEEE Latin-American Conference on Communications (LATINCOM)*, 15–17 September 2010.

[9] Zheng Wei, Liu San-yang, and QI Xiao-Gang. Integrated resources optimization in three-layer dynamic network. *JCIT: Journal of Convergence Information Technology*, 5(6):40–46, 2010.

[10] Ming Xia, Massimo Tornatore, Yi Zhang, Pulak Chowdhury, Charles Martel, and Biswanath Mukherjee. Greening the optical backbone network: A traffic engineering approach. In *Proceedings of IEEE International Conference on Communications (ICC)*, 23–27 May 2010.

[11] Yi Zhang, Pulak Chowdhury, Massimo Tornatore, and Biswanath Mukherjee. Energy efficiency in Telecom optical networks. *IEEE Communications Surveys & Tutorials*, 12(4), 2010.

[12] Steven Chamberland and Abderraouf Bahri. Designing multi-layer WDM networks with reliability constraints. In *Proceedings of Ninth International Conference on Networks (ICN)*, 11–16 April 2010.

[13] Bart Puype, Willem Vereecken, Didier Colle, Mario Pickavet, and Piet Demeester. Power reduction techniques in multilayer traffic engineering. In *Proceedings of 11th International Conference on Transparent Optical Networks (ICTON'09)*, 28 June–2 July 2009.

[14] I. Katib and D. Medhi. A network optimization model for multi-layer IP/MPLS over OTN/DWDM networks. In *Proceedings of 9th IEEE International Workshop on IP Operations & Management*, Lecture Notes in Computer Science, Vol. 5843, pp. 180–185, 2009.

[15] Stelios Androulidakis, Tilemachos Doukoglou, and George Patikis. Service differentiation and traffic engineering in IP over WDM networks. *IEEE Communications Magazine*, 46(5), May 2008.

[16] Bin Chen, Wen-De Zhong, and Sanjay K. Bose. Providing differentiated services for multi-class traffic in IP/MPLS over WDM networks. *Photon Netw. Commun.* 15:101–110, 2008, DOI 10.1007/s11107-007-0095-9.

[17] A. Varga. The OMNeT++ discrete event simulation system. In *Proceedings of the European Simulation Multiconference (ESM'2001)*, Prague, Czech Republic, June 2001.

[18] M. Vigoureux, B. Berde, L. Andersson, T. Cinkler, L. Levrau, M. Ondata, D. Colle, J. Fdez-Palacios, and M. Jäger. Multilayer traffic engineering for GMPLS-enabled networks. *IEEE Communications Magazine*, 43(7), July 2005.

[19] N. Wang, K. Ho, G. Pavlou, and M. Howarth, An overview of routing optimization for internet traffic engineering. *IEEE Commun. Surveys Tutorials*, 10(1):36–56, 2008.

[20] M. Gupta and S. Singh. Greening of the internet. In *Proceedings ACM SIGCOMM*, Karlsruhe, Germany, August 2003.

Biographies

Vassilis Foteinos graduated from the Department of Informatics Telecommunications of the National & Kapodistrian University of Athens in 2008 and received his MSc diploma from the Department of Digital Systems of the University of Piraeus in 2011. Currently he is working as a research engineer at UPRC, where he is also pursuing a PhD in traffic engineering for Future Networks.

Kostas Tsagkaris received his diploma (2000) and his PhD degree (2004) from the School of Electrical Engineering and Computer Science of the National Technical University of Athens (NTUA). He was awarded with the "Ericsson's awards of excellence in Telecommunications" for his PhD thesis. Since 2005 he is working as a senior research engineer and adjunct Lecturer in the undergraduate and postgraduate programs at the Department of Digital Systems of the University of Piraeus. He has been involved in many EU research projects including FP7/ICT UniverSelf, OneFIT and E3 and FP6/IST E2R I/II. His research interests are in the design, management and performance evaluation of wireless cognitive networks, self-organizing and autonomic networks, optimization algorithms, learning techniques and software engineering. He has published more than 100 papers in international journals and refereed conferences. He has also participated and contributed to EU and US standardization committees and working groups such as RRS and AFI-ISG groups in ETSI and IEEE DySPAN/P1900.4 group, where he has also served as Technical Editor of the published standard.

Pierre Peloso is a research engineer at Alcatel-Lucent Bell Labs in Paris area. He began his research career 12 years ago after graduating from the Ecole Nationale Supérieure de Physique de Marseille. His initial research was related to the physics and architecture of optical networks. He then worked on multilayer dimensioning. He is now focusing on autonomic networking research inside the Advanced Internet Research department, where he is part of the coordination team of the FP7 project named UniverSelf. In the meanwhile he is taking an active part in GMPLS standardization for optical networks inside IETF. He has been recognized as a Distinguished Member of the Technical Staff of Alcatel-Lucent in 2010.

Laurent Ciavaglia is currently research manager at Alcatel-Lucent Bell Labs France, in the Networking Technologies research domain, where he coordinates a team specialized in autonomic and distributed systems. Since 2010, Laurent is also leading the FP7-UNIVERSELF project (www.univerself-project.eu). In recent years, he has been working on the design, specification and evaluation of carrier-grade networks including several European research projects dealing with network control and management. Laurent is also member of the Industry Advisory Board of the FP7-Network of Excellence NESSOS. Laurent is vice-chair of the ETSI Industry Specification Group on Autonomics for Future Internet (AFI), working on the definition of standards for self-managing networks. Laurent is also participating to the IETF/IRTF as part of his activities in standardization. Laurent has co-authored more than 30 publications and holds around 30 patents in the field of telecommunication networks. Laurent also acts as member of the technical committee of several IEEE, ACM and

IFIP conferences and workshops, and as reviewer of referenced international journals and magazines.

Panagiotis Demestichas, Associate Professor, received the Diploma and the Ph.D. degrees in Electrical and Computer Engineering from the National Technical University of Athens (NTUA). From December 2007 he is Associate Professor at the University of Piraeus, in the department of Digital Systems. Most of his current research activities focus on the Information Communication Technologies (ICT) Projects, partially funded by the European Commission under the 7th Framework Programme (FP7) for research and development. More specifically, he is the Project Coordinator of the ICT OneFIT (Opportunistic Networks and Cognitive Management Systems for Efficient Application Provision in the Future Internet) Project, he serves as the deputy leader of the Unified Management Framework workpackage in the ICT UniverSelf Project and as person in charge of administrative, scientific and technical/technological aspects in the ICT ACROPOLIS (Advanced coexistence technologies for Radio Optimization in Licensed and Unlicensed Spectrum) project. Since January 2004 he has been the chairman of Working Group 6 (WG6) of WWRF, now titled "Cognitive Networks and Systems for a Wireless Future Internet" and he was the technical manager, from November 2008 until March 2010, of the "End-to-End Efficiency" (E3) project. His research interests include the design and performance evaluation of wireless and fixed broadband networks, software engineering, service and network management, algorithms and complexity theory, and queuing theory. He has several publications in these areas in international journals/magazines and refereed conference proceedings. He is a member of the IEEE, ACM and the Technical Chamber of Greece. He is the director of the M.Sc. program "Techno-economic Management and Security of Digital Systems", Department of Digital Systems, University of Piraeus, and from September 2011 has been head of the same department.

Green ICT: Self-Organization Aided Network Sharing in LTEA

Shahid Mumtaz[1], Du Yang[1], Valdmar Monteiro[1,2], Jonathan Rodriguez[1] and C. Politis[2]

[1]*Instituto de Telecomunicações, University of Aveiro, Campo Universitário, Aveiro 3810-193, Portugal; e-mail: smumtaz@av.it.pt*
[2]*WMN (Wireless Multimedia & Networking), Kingston University London, UK*

Received 29 February 2012; Accepted: 5 April 2012

Abstract

One of the targets of Green Information and Communication Technology (ICT) is to reduce the energy consumption of ICT itself. This paper provides a tutorial on a new energy reduction approach – self-organization aided network sharing, which exploits the concept of cooperation and adaptation in the cellular system level. Detail reviews from different perspectives including the academic literature, the existing standard, etc. are provided. Moreover, a simply example with simulation results is provided to demonstrate the energy efficiency of the proposed scheme.

Keywords: network sharing, self-organization, energy efficiency, LTEA.

1 Introduction

Energy efficiency and low carbon strategies have attracted a lot of concern. The goal for 20% energy efficiency and carbon reduction by 2020 drove the Information Communication Technologies (ICT) sector to strategies that incorporate modern designs for a low carbon and sustainable growth [1, 2]. The ICT sector is part of the 2020 goal and participates in three different ways. In a direct way, ICT are called to reduce their own energy demands (green

Journal of Green Engineering, Vol. 2, 215–232.

networks, green IT), in an indirect way ICT are used for carbon displacements and in the systematic way ICT collaborate with other sectors of the economy to provide energy efficiency (smart-grids, smart buildings, intelligent transportations systems, etc.). As described in [3] green ICT is defined as: *Use of ICT for optimizing societal activities in order to improve environmental sustainability.*

This paper is focused on the direct way, which is to reduce the energy consumption of ICT itself, especially the energy consumption of wireless cellular network. Throughout the global community, wireless communications have had a profound social economic impact, enriching our daily lives with a plethora of services from media entertainment to more sensitive applications such as e-commerce. Looking towards the future, although voice and SMS are still major sources of revenue, mobile traffic will account for a large chunk of the internet highway. To cope with this increased demand, operators are required to invest more in core infrastructure, and deploy more advanced technologies. In fact, already in today's market operators are deploying over 120,000 new Base Stations (BSs) on a yearly basis across the world. Moreover, the mobile technologies is fast evolving from the 3rd Generation (3G) supporting 384 kb/s downlink in 2001 to the Long Term Evolution (LTE) supporting 300 Mb/s downlink in 2010. This development of wireless communication brings also the ever increasing energy consumption. For example, a medium sized cellular network uses as much energy as 170,000 homes. While the cost of powering the existing BSs accounts for a staggering 50% of a service provider's overall expenses. Therefore new solutions are required whereby operators can accommodate this additional traffic volume whilst reducing their investment in new infrastructure, and beyond that significantly reduce their energy bill.

In order to reduce the energy consumption, cooperation and adaptive optimization are two approaches often employed. A well-known cooperation example is that the cell-edge users transmit their information to the BS through relays located closer to the cell-center. Adaptive technology has been well studied in the link-level, which is specifically the adaptive modulation and coding scheme for optimizing the spectral efficiency so as to save energy. It has also been studied in the MAC-level such as the multi-user scheduling schemes so as to optimize the channels shared by multiple users. This paper is aimed at providing a tutorial on self-organization aided network sharing for energy efficiency, which extends the cooperation concept into network operators, and extends the adaptive optimization concept into the entire network resource management.

Network sharing is one of the most attractive solutions for energy and other costs reduction, in which operators cooperate with each other by sharing infrastructure, operational functions and even risks in a bid to reduce the capital and operational expenditure of the network. It is claimed that network sharing is able to achieve up to 65% saving in both roll-out capital expenditure (CAPEX) and network operations-plus-maintenance expenditure (OPEX) [4]. As a result, the regulators over the world are now encouraging this approach, although some restrictions still remain in order to avoid monopolies. Network sharing has already been implemented in various countries such as India, UK and USA. However, despite the recent efforts, the current state-of-the-art on network sharing is still in its infancy. In practice, the widely implemented network sharing is currently deployed in a very simple form of so-called passive sharing (e.g., sharing the towers), while active sharing and roaming-based sharing, which can provide more significant energy and cost saving, needs to be further investigated to support seamless and dynamic sharing. In addition, from a business point of view, the lack of clear cooperation strategies can hinder even passive network sharing among competing operators, which demonstrate the requirement of proper network sharing business models for providing incentives to the stakeholders. Hence, there is a need for further research to address all aspects of network sharing to achieve its full potential in cost and energy per bit reduction.

Once a sharing agreement is made among network operators, an effective management of the resultant network becomes crucial factor for the business success, and it is much more complex since it includes joint planning of the footprints, configuring the network parameter, etc. Self-Organizing Network (SON) techniques are capable of providing cost-effective network management for the complex shared network.

The rest of this tutorial paper is organized as follows. Section 2 provides a review on network sharing. Section 3 provides a review on self-organization network. Section 4 provides a simple example with simulation results, which demonstrated a significant gain in energy saving. A conclusion is presented in Section 5.

2 Network Sharing

From technical perspective, several architecture-level approaches have been proposed by vendors (e.g. Ericsson [5] and Nokia [6]), as well as the EU competition Directorate [7]. Although the proposed approaches from different organizations are not identical, they can be generally categorized into three clusters:

- *Passive network sharing* in which operators share network assets that are not considered to be an "active" part of providing services, such as the sites and civil engineering elements (towers, shelters, air conditioning and cooling systems, AC and DC power supply, diesel generators).
- *Active network sharing* where operators share BS elements like the Radio Frequency (RF) chains, antennas or even Radio Network Controllers (RNC).
- *Roaming-based network sharing* where one operator relies on another operator's coverage on a permanent basis.

Figure 1 illustrates a general architecture of a wireless system and six different network sharing approaches. In general, the more network assets and operational functions are shared, the more cost savings are obtained accompanied by the loss of control of the entire network [8]. For example, in the case of full sharing, operator A relies on the operator B's network assets and operation functions to provide service to customers in operator B's coverage area. Both investment and operational cost are shared between A and B, while operator A has no control of the service quality at all. Apart from these architecture-level approaches, studies considering lower-layer technologies [9], such as Medium Access Control (MAC)/Physical (PHY) layer, in network sharing are very limited.

From the business perspective, the analysis results from both the academia and industry have demonstrated the benefits of network sharing in terms of meeting the increasing data demand and reducing the network cost [10]. Moreover, network sharing will probably further break down the value-chain, cultivate new business specialize in a certain area such as building tower, and create mobile virtual network operators who own no other physical assets except their home location resistor and billing system. In addition, the adoption of the aforementioned technical approaches depends on the business situation. For example, a high-degree network sharing would not be successful if two operators cannot agree with the technology updating plan because of different business strategies.

The benefits shown in technical and business analysis change the regulators' attitude from resistant to welcoming network sharing, although the tension still exists, and some resource sharing such as frequency pool are still not allowed. A summary of the policies in some EU countries are listed in [8]. In practice, the starting point is usually the sharing of sites, includ-

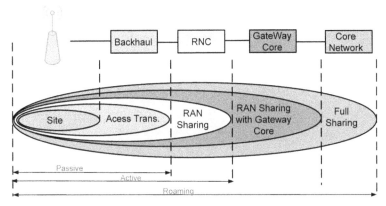

Figure 1   General illustration of wireless network architecture and different degrees of the network sharing.

ing passive infrastructure, as is the case of the recently announced joint venture Indus Towers, into which Indian operators Vodafone Essar, Bharti Infratel, and Idea Cellular are planning to merge their approximately 70,000 existing sites, and which should be responsible for the further network roll-out. Where emerging/developing market operators are looking at economic option for coverage and capacity growth, operators in mature markets are seeking cost optimization and technology refresh, like UK operators Orange and Vodafone, where active sharing are taken into consideration. Besides these successful examples, many Network Sharing deals have failed (despite considerable efforts from the operators' side). According to Ericson's analysis , common reasons for failed partnerships are: (1) the lack of confidentiality, trust and asset valuation; and (2) the complexity in RAN Sharing (Decision/Execution) often offsets savings.

2.1 Network Sharing in LTE-A Standard

3GPP has specified network sharing architectures shown in Figures 2 and 3, which allow a singular physical Universal Terrestrial Radio Access Network (UTRAN)[1] deployment to be shared between multiple Core Network (CN) operators, each with their own separate CN infrastructure deployments. Two architectural variations of Network Sharing are defined:

---

[1] UTRAN is the RAN employed in LTE system. Its evolution version E-UTRAN is employed in LTE-A system.

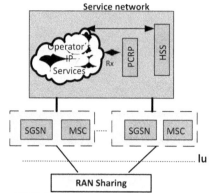

Figure 2 MOCN: Multiple Operator Core Network.

- MOCN: Multiple Operator Core Network.
- GWCN: Gateway Core Network.

This architecture applies to UTRAN and E-UTRAN (i.e. not GERAN: GSM/EDGE Radio Access Network). While Release 6 UE devices are required to fully exploit this Release 6 network capability, pre-Release 6 UEs are also supported by this architecture. MOCN allows a single UTRAN to be directly connected to up to five separate CN operators, thereby offering the opportunity for a new CN Operator to avoid RAN deployment. A separate Iu interface (PS and CS) is deployed for each CN operator. GWCN also allows a single UTRAN to be used by multiple operators

3 Self-Optimization Network

The appearance of SON algorithms represents a continuation of the natural evolution of wireless networks, where automated processes are simply extending their scope from just frequency planning to overall network resource management. The rationale for SON automation can be grouped into two broad categories:

1. Previously manual processes that are automated primarily to reduce the manual intervention in network operations in order to obtain operational and/or deployment savings. Automating repetitive processes clearly

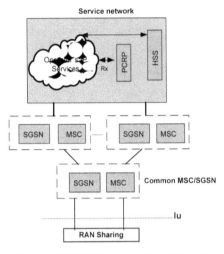

Figure 3 GWCN: Gateway Core Network.

saves time and reduces effort. Auto-configuration and self-configuration fall into this category.

2. Processes that require automation because they are too fast, too granular (per-user, per-application, per-flow, as a function of time or loading), and/or too complex for manual intervention. Automatically collected measurements from multiple sources (e.g., from user devices, individual network elements, and on an end-to-end basis from advanced monitoring tools) will provide accurate real-time and near real-time data upon which these algorithms can operate thus providing performance, quality, and/or operational benefits.

Consequently, substantial opportunities exist for cross-layer, end-to-end, and per-user/per-application/per-flow optimizations for extracting additional performance benefits and management flexibility.

In one of Next Generation Mobile Network (NGMN) white papers high-level requirements for Self-Optimization network strategy were included with set of use cases defined, covering multiple aspects of the network operations including planning, deployment, optimization and maintenance. More explicitly, some of the use cases were:

1. Plug & Play Installation.
2. Automatic Neighbour Relation Configuration.
3. OSS Integration.

4. Handover Optimization.
5. Minimization of Drive Tests.
6. Cell Outage Compensation.
7. Load Balancing.
8. Energy Savings.
9. Interaction Home/Macro BTS.
10. QoS Optimization.

Most of the use cases are addressed in 3GPP. 3GPP initiated the work towards standardizing self-optimizing and self-organizing capabilities for LTE, in Release 8 and 9. The standards provide network intelligence, automation and network management features in order to automate the configuration and optimization of wireless networks to adapt to varying radio channel conditions, thereby lowering costs, improving network performance and flexibility. This effort has continued in Release 10 with additional enhancements in each of the above areas and new areas allowing for inter-radio access technology operation, enhanced Inter-Cell Interference Coordination (eICIC) [14], coverage and capacity optimization, energy efficiency and minimization of operational expenses through Minimization of Drive Tests (MDT) [14].

4 Simulation Results and Discussion

4.1 Simulation Scenario

As shown in Figure 4, we considered a common area covered by *operator_1* (red) and *operator_2* (green). Assuming that *operator_1* has more subscribers in this area than *operator_2*, this area is covered by two small cells – *cell_1* and *cell_2* – of *operator_1*, and covered by one large cell named *cell_3* of *operator_2*. Without any sharing, users are only able to communicate to the BSs belonging to the operator they subscribed to. Hence, the edge-user *user_1* and *user_2* shown in Figure 4 suffer a relatively large path-loss, which results in high transmit power requirement and low cell-edge throughput.

The situation of employing network sharing is illustrated in Figure 5. The same geographic area shown in Figure 4 is considered. We assumed that *operator_1* and *operator_2* have service-level agreement. Moreover, we assumed that every user is capable of working on both operators' frequency bands. Therefore, the users are allowed to connect to BSs belonging to any operator. As a result, *user_1* and *user_2* will be able to connect to the nearest BS so as to save energy. Referring to the network sharing categories described in Section 2, the scenario we considered is *an active RAN sharing*, whose energy

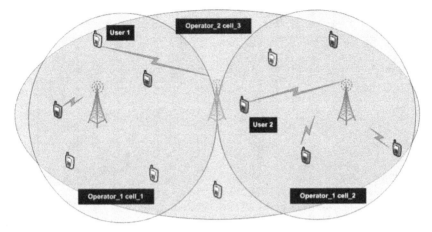

Figure 4 A common area covered by two independent operators: operator_1 (red) with cell_1 and cell_2; operator_2 (green) cell_3.

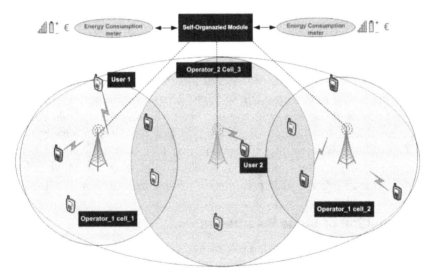

Figure 5 A common area covered by two service-share operators aided by self-organization.

saving will be investigated in this section. Moreover, the two operators could either further share or not share their MSC, core network, etc., depending on their policies.

Furthermore, a self-organizing optimization module is employed at every BS, which is capable of collecting reference signal received power (RSRP)

Table 1 Simulation parameters

Parameters	Downlink LTEA (FDD) System
Carrier frequency f_c	Operator_1: 2 GHz
	Operator_2: 1.9 GHz
Bandwidth	10 MHz
Fast fading model	Rayleigh fading using Pedestrian B model (6 taps, SISO) Urban
Number of cells	Hexagonal grid, 19 three-sectored cells
Number of users	100 per cell
User speed	3 & 30 km/h
Users power	23 dBm
BS transmit power	43 dBm
Inter-site distance	500 m
Time transmission interval (T_{ti})	1 ms (sub-frame)
Number of resource block	50 RB in each slot, 7 symbol, number of subcarriers per RB=12, total subcarrier=600
Link adaptation	EESM (Exp Effective SINR Mapping)
Traffic model	Mix Traffic Option (VoIP & NRTV), Cell Arrival Rate: 10 user/cell/sec
Radio resource management	RR, PF, Max C/I
Number of MCS	12 (from QPSK 113 to 64-QAM 3/4)

and Signal-to-Noise-plus-Interference-Ratio (SINR) from every user, monitoring the traffic load at the BS, so as to accordingly adjust the downlink transmit power at the BSs, with the purpose of finding the optimal coverage range, enhancing the throughput, and minimizing the energy consumption.

Throughout this section, we assume that LTEA-FDD technologies are deployed by all operators and cells. A hundred users are uniformly distributed over the area of interest. Other simulation parameters are shown in Table 1.

4.2 Definition of Some Parameters

In this subsection, we define the following parameters mostly related to self-organizing algorithm.

- *Coverage area of each operator*
  We assumed that the coverage area has a hexagonal shape, and this symbol r_{opi} represents the edge length of the hexagonal area covered by *operator_i*. As a result, the area covered by *operator_i* is calculated as

$$\varphi(op_i) = \frac{3\sqrt{3}}{2} r_{opi}^2.$$

Moreover, the length of the hexagon changes depending on the transmission power P_c.

- *Traffic load*

The traffic load is a metric that represents the entire traffic load in the area covered by r_{opi}. That is:

$$T_{opi} = \sum_{x=1}^{n} \sum_{y=1}^{m} p(x, y) \cdot \delta_{opi}(x, y),\tag{1}$$

where the symbol $p_{poi}(x, y)$ is the traffic load at point (x, y), and $\delta(x, y) \in \{0, 1\}$ is the function that indicates where that point (x, y) is covered ($\delta(x, y) = 1$) or not ($\delta(x, y) = 0$) by operator i.

- *Operators sharing area*

This metric represents the ratio of the sharing area between *operator_i* and other operators. That is:

$$SA_{opi} = \frac{\sum_{j \neq i}^{N} \varphi(op_i, op_j)}{\varphi(op_i)},\tag{2}$$

where $\varphi(op_i, op_j)$ is the area value of sharing region between op_i and op_j. The symbol SA_{opi} represents the sharing ratio. Large SA means large sharing area. Index j is the operator index, N is total number of operators, and $\varphi(op_i) = \frac{3\sqrt{3}}{2} r_{opi}^2$ is the area value of the region covered by *operator_i*.

- *Energy*

Energy of each operator in sharing region is measured in Joule/Bit:

$$\text{Energy Efficiency} \sim \frac{\text{Joule}}{\text{bit}}\tag{3}$$

- *Reference Signal Received Power (RSRP)*

Supposed that transmit power of the BS in *cell_i* is P_i in dBm, the pathloss from this BS to a user *ue* equal to $L_{i,ue}^{p}$ (dBm), the corresponding shadow fading $L_{i,us}^{s}$ (dBm) having a log-normal distribution with standard deviation of 3 dB, and the fast fading is represented as $L_{i,ue}^{f}$ (dBm), the RSPR in dBm between this user and the BS in *cell_i* is formulated as

$$\text{RSRP}_{i,ue} = P_i - L_{i,ue}^{p} - L_{i,ue}^{s} - L_{i,ue}^{f}.$$

- *Signal-to-Interference-and-Noise-Ratio (SINR)*

Supposed that the user *ue* is connected to the BS in *cell_i*, and received interference from other cells, the SINR is formulated as

$$\text{SINR}_{ue} = \frac{\text{RSPR}_{i,ue}\ (\text{mW})}{\sum_{j \neq i}^{N} \text{RSPR}_{j,ue}\ (\text{mW})} + N_0\ (\text{mW}),$$

where all values are converted from dBm into mW, N is the total number of cells, and the symbol N_0 represents the noise power.

4.3 Simulation Results

In this section, several performance metrics including throughput, SINR distribution, and energy consumption of the *operator_1 cell_1* are measured in three situations:

1. without network sharing as shown in Figure 4;
2. with network sharing and constant transmit power as shown in Figure 5;
3. with network sharing and SON aided transmit power adaption as shown in Figure 5.

4.3.1 Network Sharing

Scenario definition is shown in Figure 5. Figure 6 shows the simulation results of average cell-edge user throughput versus the entire cell load of *operator_1 cell_1*. Traffic load and sharing region of each operator is calculated according to equations (1) and (2). A user is defined as cell-edge user if the distance between this user and its served BS larger than 2/3 radius of cell. Proportional fair scheduling is employed. It is demonstrated that the achievable average throughput of cell-edge users with network sharing is significantly higher than the one without sharing strategies. For instance, cell-edge average user throughput gain in non-sharing scheme is 33% at the cell load of 12 Mb/s, while in multi-operator sharing scheme the average user throughput is increased approximately by about 22% on same cell load. This is because that aided with network sharing, the cell-edge users such as *user_1* shown in Figure 6 become able to communicate with a nearer BS. As a result, the channel quality as well as the achievable throughput experienced by the cell-edge user is improved.

4.3.2 SON-Based Network Sharing

As illustrated in Figure 7, the self-organizing module takes input parameter the RSRPs and SINRs from users' feedback, as well as the current traffic load from the BS counter. It adjust the transmit power according to these

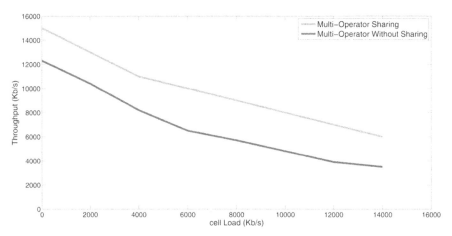

Figure 6 Averaged throughput of cell-edge op1 users versus the total cell load, when (1) op1 and op2 has sharing agreement, and (2) op1 and op2 has no sharing agreement.

parameters in order to maximizing the SINR values of the active users. For example, when the demanded traffic load increase in the *operator_1 cell_1* area, and is over the maximum affordable traffic load. Users who are in the coverage area of *operator_1 cell_1* cannot be served normally because of traffic overloading. The SON module detects this situation by observing a high ratio of admission denies in *cell_1*. As a consequence, the SON module adjust the transmit power at *cell_1* and *cell_3*. More explicitly, the BS transmit power at cell_1 is reduced, and the BS transmit power at *cell_3* is increased; in order to handover the cell-edge users' requirement from *cell_1* to *cell_3*.For simulation purposes, fixed number of users are deployed in each operator to see the performance of SON module.

Figure 8 show the Cumulative Density Function (CDF) of the users' SINR in cell_1. With SON based network sharing almost 85% of users achieve the required minimum SINR value 10dB. Without SON based network sharing, only 75% of users get required SINR value. This gain is achieved due to the optimal power allocation of SON module. SON module adapted its power according to traffic load in each operator cell.

4.3.3 Energy Efficiency
Energy efficiency is calculated according to equation (3). Figure 9 shows the energy consumption comparison of SON based Network sharing (SNS), simple network sharing and no sharing in term of energy per packet. Packet

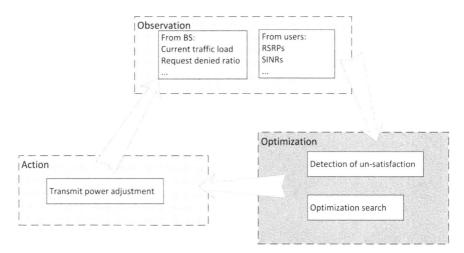

Figure 7 Flow chart illustration of the iterative optimization processing of SON module.

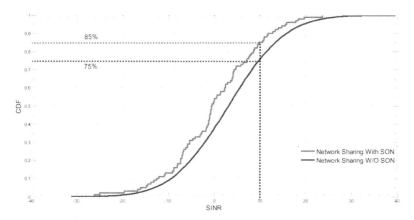

Figure 8 CDF of SON based network sharing.

size depends upon AMC scheme on each TTI. Mobile users are randomly distributed each operators site. We change the maximum allowed energy consumption per packet, and draw the overall energy consumption as shown in Figure 9, where we see clearly that compared with network sharing scheme, SNS achieves significant energy savings without network sharing yield minimum energy saving.

Figure 9 Energy efficiency comparisons.

5 Conclusions

ICT brings its intelligence to diverse networks and telecommunications industry. As these networks become more intelligent and complex, new solutions for network management are needed. ICT enables network optimization on multiple levels; energy consumption, throughput etc. SON has potential to self-configuring, self-optimizing and self-healing and minimizes the energy consumption in the network. In this paper we provided a tutorial on a new energy reduction approach – self-organization aided network sharing. We considered self-optimizing property of SON and showed from a simply example that SON based network sharing, optimized the system performance and minimized energy consumption in multi-operator scenario.

References

[1] International Telecommunication Union (ITU), Report on Climate Change, October 2008.
[2] G. Koutitas and P. Demestichas. A review of energy efficiency in telecommunication networks. In *Proceedings in Telecomm. Forum (TELFOR)*, Serbia, November, pp. 1–4, 2009.
[3] Jussi Ahola, Toni Ahlqvist, Miikka Ermes, Jouko Myllyoja, and Juha Savola. ICT for Environmental Sustainability. Green ICT Roadmap. Espoo 2010. VTT Tiedotteita – Research Notes 2532, 51 pp., 2010.
[4] S. Zehle and G. Friend. Network sharing business plan. *Network*, 1–34, July 2010.
[5] Ericsson. Shared networks: An operator alternative to reduce initial investments, increase coverage and reduce Time to Market for WCDMA, pp. 1–9, 2003.

 [6] Nokia. Network sharing in 3G. 2001.

 [7] Northstream. 3G Rollout Status. 2002.

 [8] T. Frisanco, P. Tafertshofer, P. Lurin, and R. Ang. Infrastructure sharing and shared operations for mobile network operators From a deployment and operations view. In *Proceedings of IEEE Network Operations and Management Symposium*, pp. 129–136, 2008.

 [9] S.A. Al Qahtani and A.S. Mahmoud. Simulation based study of adaptive rate scheduling for multi-operator 3G mobile wireless networks. In *Proceedings of IEEE 65th Vehicular Technology Conference*, pp. 1209–1213, 2007.

[10] C. Beckman and G. Smith. Shared networks: Making wireless communication affordable. *IEEE Wireless Communications*, 79–85, April 2005.

[11] L. Nielsen. Network sharing: Technical possibilities in GSM/UMTS and LTE, pp. 1–10, 2010.

[12] H. Hu, J. Zhang, X. Zheng, Y. Yang, and P. Wu. Self-configuration and self-optimization for LTE networks. *IEEE Communications Magazine*, 48(2):94–100, 2010.

[13] J. Boggis, J. Castro, H. Liu, A. Minokuchi, D. Salam, P. Schwinghammer, and M. Shahbaz. Next generation mobile networks beyond HSPA & EVDO. A white paper by the NGMN Alliance, 5 December 2006.

[14] Nokia Siemens Networks LTE-Advanced. The advanced LTE toolbox for more efficient delivery of better user experience. Technical white paper, LTE-Advanced.

[15] Sin-Seok Seo, Sung-Su Kim, N. Agoulmine, and J.W. Hong. On achieving self-organization in mobile WiMAX network. In *Proceedings of Network Operations and Management Symposium Workshops (NOMS Wksps)*, 2010 IEEE/IFIP Issue, 19–23 April 2010.

Biographies

Shahid Mumtaz received his MSc. degree from the Blekinge Institute of Technology, Sweden and his Ph.D. degree from University of Aveiro, Portugal. He is now a senior research engineer at the Instituto de Telecomunicações – Pólo de Aveiro, Portugal, working in EU funded projects. His research interests include MIMO techniques, multi-hop relaying communication, cooperative techniques, cognitive radios, game theory, energy efficient framework for 4G, position information assisted communication, joint PHY and MAC layer optimization in LTE standard. He is author of several conference, journal and book chapter publications.

Jonathan Rodriguez received his Masters degree in Electronic and Electrical Engineering and Ph.D from the University of Surrey (UK), in 1998 and 2004 respectively. In 2002, he became a Research Fellow at the Centre for Communication Systems Research at Surrey and responsible for managing the system level research component in the IST MATRICE,

4MORE and MAGNET projects. Since 2005, he is a Senior Researcher at the Instituto de Telecomunicações – Aveiro (Portugal), where he is leading the 4TELL Wireless Communication Research Group. His research interests include Radio Access Networks for current and beyond 3G systems with specific emphasis on Radio Resource Management, Digital Signal Processing and PHY/MAC optimization strategies.

Valdemar Monteiro received his *Licenciatura** and Masters degree in Electronic and Telecommunications from the University of Aveiro (Portugal), in 1999 and 2005 respectively. In 2000, after his graduation, he became a Research Fellow at Instituto de Telecomunicações – Aveiro and has worked for international research projects that include IST SAMBA, IST MATRICE, 4MORE and UNITE. In 2008 he joined CV Movel (Cabo Verde), Cape Verde main Mobile Operator to work as Switch Engineer. Since March 2009 he is working for Instituto de Telecomunicações on a PhD programme. He is the author of several conference and journal publications, and has carried out consultancy for operators (Portugal Telecom Inovação) and HSDPA standardisation. His research interests include Radio Access Networks for legacy and beyond3G systems with specific emphasis on IP networking, Cooperative Radio Resource Management and PHY/MAC optimisation strategies.

Du Yang received her BEng. degree from the Beijing University of Posts and Telecommunications (China) in 2005; and her MSc. and Ph.D. degrees from University of Southampton (UK), in 2006 and 2010 respectively. She was a recipient of the Mobile VCE Scholarship. She is now a Post-doctoral researcher at the Instituto de Telecomunicações – Pólo de Aveiro, Portugal, working in the EU funded WHERE2 project. Her research interests include MIMO techniques, multi-hop relaying communication, position information assisted communication, joint PHY and MAC layer optimization in LTE standard.

Christos Politis is a Reader (Associate Professor) in Wireless Communications at Kingston University London, School of Computing & Information Systems (CIS), Faculty of Science, Engineering and Computing (SEC). There he leads a research team on Wireless Multimedia & Networking (WMN) and teaches modules related to communications. Christos is the course director for the 'Wireless Communications/ WC', 'Networks and Data Communications/ NDC' and 'Networks and Information Security/NIS' postgraduate taught courses. He holds two patents and has published more than 120 papers in international journals and conferences proceedings and chapters in four books. Christos was born in Athens, Greece and holds a PhD and MSc from the University of Surrey, UK and a B.Eng. from the Technical University of Athens, Greece. He is a senior member of the IEEE, a member of IET and a member of Technical Chamber of Greece.

Achieving Energy Efficiency through the Opportunistic Exploitation of Resources of Infrastructures Comprising Cells of Various Sizes

Dimitrios Karvounas, Andreas Georgakopoulos, Dimitra Panagiotou, Vera Stavroulaki, Kostas Tsagkaris and Panagiotis Demestichas

Department of Digital Systems, University of Piraeus, 80 Karaoli & Dimitriou str, Piraeus 18534, Greece; e-mail: dkarvoyn@unipi.gr

Received 4 March 2012; Accepted: 6 April 2012

Abstract

One of the main challenges of the arising wireless world is to support a wide set of QoS-demanding services. However, the solutions that will be considered should not only take into consideration the optimization of the network performance, but should also be energy efficient both for the operator and the end-user. This paper presents the concept of Opportunistic Networks (ONs) as an energy efficient method to exploit wireless networks. The cases studied in this work comprise a macro base station and femtocells. The proposed solution will offload a proportion of the traffic of the base station to the femtocells, through the creation of an ON. Therefore, the base station will consume less energy since terminals will be rerouted to the femtocells. In addition, the femto-terminals will operate to lower power levels leading to battery savings. Results from simulations are provided to confirm the proposed solution.

Keywords: opportunistic networks, capacity extension, future internet, femtocells, energy consumption.

Journal of Green Engineering, Vol. 2, 233–253.

1 Introduction

The wireless world is at the centre stage due to the need of people for mobility in conjunction with a wide set of applications. However, current networks were not designed for the usage level that faces today. Therefore, there is need to enhance networks to handle the increased traffic and offer adequate services in terms of quality of service (QoS), security and mobility. Until recently the provided solutions were not taking into account energy aspects. Nonetheless, in recent years research has turned to solutions that provide efficiency with respect to economic, societal and energy criteria.

A common problem that network operators face is that at any moment their network may experience capacity problems due to a plethora of reasons: an increase in the amount of customers, e.g. due to a mass event, a displacement of users, a malfunction in the infrastructure, etc. As a result, congestion issues arise. However, the solution that will be adopted should not only take into consideration network optimization, but should also exploit green technology that will result in capital and operational expenditure (CAPEX and OPEX respectively) savings.

In order to confront this problematic situation, this work will focus on the use of Opportunistic Networks (ONs) that seems to be a promising solution for the problem of the capacity extension of congested infrastructure (e.g. base stations (BSs)). ONs are operator-governed, coordinated extensions of the infrastructure and are created dynamically for a limited time [1]. The life cycle of an ON consists of four phases: suitability determination, creation, maintenance and termination. At the suitability determination phase, the suitability of an ON at a specific time and area is determined. The final configuration of an ON is done in the creation phase. The selection of aspects such as involved nodes, used spectrum for the ON are done based on context information and more accurate estimations of the radio environment. At the maintenance phase monitoring will be done to ensure that the ON is still valid and efficient for what it was created for. According to the monitoring information reconfiguration or termination functionalities can be triggered. Reconfiguration will ensure the most efficient operation of the ON. When the ON operation is no longer necessary or suitable, the ON will be terminated [2].

The proposed solution to relieve the congested infrastructure is to exploit the opportunity that femtocells offer, e.g. extra capacity, and create an ON in order to offload a proportion of the traffic to alternate Wireless Access Networks (WANs) that are not problematic. However, optimization of the network performance is not the only target, but energy efficiency constitutes the prominent target.

In our work a network is considered that consists of a macro BS, terminals that are served via the macro BS and femtocells that are distributed within the coverage area of the macro BS. This paper will focus on the benefits that derive from the exploitation of femtocells that lead to energy savings both for the infrastructure and the end-user side.

The rest of the paper is structured as follows: Section 2 provides state of the art work related to our problem. Section 3 provides an overview of the capacity extension through femtocells concept. Section 4 illustrates the mathematical formulation that was used for the problem. Simulation methodology and results are provided in Section 5 and, finally, the paper concludes with Section 6.

2 Related Work

This section provides a discussion of state-of-the-art work related to the examined problem.

Mobile data offloading is the use of complementary networks for the delivery of data that was originally targeted for cellular networks [3]. The main Radio Access Technology (RAT) used for the data offloading is Wi-Fi. The main advantage of Wi-Fi is that it operates at the unlicensed spectrum and therefore causes no interference to cellular networks. In addition, in most urban areas Wi-Fi is ubiquitous available deployed by operators or by single users. In [4] the authors evaluate the potential costs and gains for providing Wi-Fi offloading in a metropolitan area by using large scale real mobility traces for empirical emulation.

Virtual carrier is a method to exploit the cooperation of licenced (e.g. UMTS, LTE) and unlicensed (Wi-Fi) spectrum to improve network capacity. The levels of synergy vary, depending on whether a multiprotocol client connects to distinct 3G/4G and Wi-Fi Access Points (APs) or to integrated 3G/4G/Wi-Fi devices that can implement cooperative utilization of the available spectrum. In this case, an additional/virtual carrier is available to increase network performance [5]. In [6] the authors provide further information for multiradio interworking between Worldwide Interoperability for Microwave Access (WiMAX) and Wi-Fi.

The authors in [7] propose a distributed solution for designing the radio resource allocation of downlink transmissions in femtocell networks. The network to which the algorithm is evaluated consists also of a macro BS and several femtocells deployed within its area. Furthermore, the authors in propose an optimal power allocation strategy based on modelling the interferer's activity as a two-state Markov chain. In the single femto-user access is shown how to maximize the expected value of femto-user rate, averaged over the interference statistical model. Then, the approach is extended to the multiuser case, adopting a game-theoretic formulation to devise decentralized access strategies, particularly suitable in view of potential massive deployment of femtocells. Moreover, in [9] the authors propose a distributed resource allocation strategy for the access of opportunistic users in cognitive networks. The solution is based on a social forage swarming model, where the search for the most appropriate slots is modelled as the motion of a swarm of agents in the resource domain, looking for "forage", representing a function inversely proportional to the interference level.

In addition, in [10] the authors suggest a distributed inter-cell power allocation algorithm where each cell computes by an iterative process its minimum power budget to meet its local QoS constraints. Their results show how their work permits to reduce both transmission power and harmful effects of in-band interference, while meeting QoS constraints of users in each cell. Moreover, the authors in [11] focus on a power control mechanism and propose a novel approach for dynamic adaptation of femtocells' transmitting power. The basic idea is to adapt the transmitting power of femtocells according to current traffic load and signal quality between user equipments and the femtocell in order to fully utilize radio resources allocated to the femtocell.

Furthermore, in [12] the EARTH energy efficiency evaluation framework (E^3F) evaluates the performance of a Radio Access Network (RAN) at system level using multi-cell system information. Also, it introduces a BS power model that is used to monitor the network energy consumption. In addition, relevant performance metrics related to the energy consumption of the network are defined and used as a complement to the classical key performance indicators. Moreover, the authors in [13] present new concepts to save energy in small-cell wireless communication BSs. The power consumption of these BSs is dominated by three components: the digital baseband engine, the analogue RF transceiver and the power amplifier. For these components, energy adaptation solutions are identified and quantified in function of the signal load.

3 Scenario Overview

As mentioned in the previous sections, an operator may face at any moment congestion in the infrastructure. In order to overcome this problematic situation our approach proposes the use of ONs for the infrastructure capacity extension through femtocells. Femtocells are fully featured but low power BSs that operate in licenced spectrum in order to connect mobile terminals to the network of an operator through residential broadband connections [14]. Therefore, the signal strength improves significantly resulting to better QoS. Moreover, there are different ways for a user to access a femtocell. The Closed Subscriber Group (CSG) model, where only a limited number of users have access to femtocell's resources. These users are defined by the femtocell owner. The Open Subscriber Group (OSG) where all operator's customers have access to the femtocell. The Hybrid Subscriber Group (HSG) where a limited amount of femtocell's resources are available to all users, while the rest are operated in a CSG manner [15].

In the capacity extension problem, as soon as a BS experiences congestion issues the Network Management Entity (NME) is notified in order to trigger the solution procedure. If neighbour femtocells exist, the suitability determination phase will define the terminals that are able to be redirected to femtocells, i.e. terminals with low mobility level. These terminals will be rerouted to femtocells relieving the congested BS, while the remaining to the BS terminals will be able to experience higher QoS. In addition, due to the fact that a proportion of terminals will be assigned to the femtocells, the BS will no longer be connected to them leading to a reduction in its energy consumption and savings in OPEX. Furthermore, the terminals that are served by the femtocells need lower transmission power to communicate with them and as a result their battery lifetime is expected to increase.

The benefits that derive from the concept of the capacity extension through femtocells are depicted in Figure 1.

4 Mathematical Formulation

For the solution we propose, a network that comprises a macro BS, femtocells and terminals is considered. The target is to estimate the energy consumption of the BS in relation with the connections with the terminals that it serves. For this purpose the Okumura–Hata propagation model for suburban areas is introduced [16] in order to enable us to calculate the transmission power that is needed for the communication between the BS and its terminals. Moreover, it

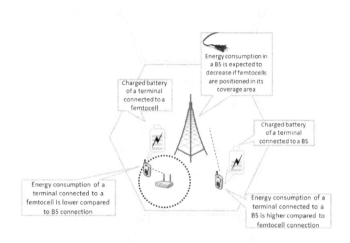

Figure 1 The benefits of the capacity extension through femtocells.

is assumed that not all terminals are transmitting data simultaneously. There-fore, if T is the set of terminals, $IT \subseteq T$ will denote the set of idle terminals and $AT \subseteq T$ will denote the set of active terminals, i.e. the set of terminals that are transmitting data. Apparently $IT \cup AT = T$. The transmission power of the BS for a specific time t can be estimated as follows:

$$P(t) = \sum_{i \in IT} P_i + \sum_{a \in AT} P_a, \tag{1}$$

where P_i is the power needed to communicate with an idle terminal and P_a is the power needed to communicate with an active terminal. P_a can be estimated via the Okumura-Hata propagation model, while P_i is assumed to be a small proportion of the transmission power that the BS would need to communicate with the terminal if it was active.

The energy consumption of the BS until a specific time t_0 can be calculated as

$$C(t_0) = \sum_{t \leq t_0} P(t) \cdot t, \tag{2}$$

Table 1 Simulation parameters.

Parameter name	Parameter value
BS antenna height	20 m
Terminal's antenna height	1.6 m
Frequency of transmission	2000 MHz
Terminal's sensitivity	−120 dBm

where t is the amount of time for which the BS was transmitting at power level $P(t)$.

Finally, the propagation model that was used for the femtocells is the one described in [17].

5 Simulation Results

5.1 Methodology

In order to evaluate the operation of the proposed concept, 3 test cases are considered for estimating the power consumption of the macro BS. The values of the parameters that were considered during the simulations are depicted in Table 1.

All test cases were simulated by using a customized version of the Opportunistic Network Environment (ONE) simulator [18] in a system with Intel Pentium D CPU at 2.80 GHz and 2.5 GB of RAM.

The topology that is used for the evaluation of the proposed solution consists of a macro BS, 30 femtocells that are located in a uniform distribution within the area of the BS and 50 terminals that move randomly within the area of the BS with an average velocity of 1 m/s.

5.2 Test Cases and Results

Test case 1 illustrates the impact of enabling femtocells to the network, to the power consumption of the macro BS, while test case 2 aims at showing how the number of users affects the power consumption of a macro BS. Finally, test case 3 depicts the impact of the use of femtocells to the battery lifetime of a terminal.

These test cases are indicative scenarios that highlight the benefits that derive from the concept of the capacity expansion through femtocells.

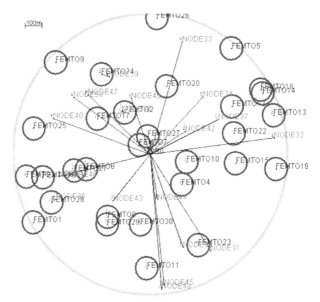

Figure 2   A macro BS with femtocells within its coverage.

5.2.1 Test Case 1

In this test case, the impact of the number of femtocells on the power consumption of the BS will be examined. Therefore, a macro BS area is considered as shown in Figure 2. The outer circle depicts the coverage of the macro BS, while the smaller circles indicate the coverage of the femtocells. Firstly, the case where the BS operates without nearby femtocells will be studied and then 10 femtocells will be enabled each time. The simulation time was 2000 secs.

Figure 3 depicts a snapshot from the case where all terminals are connected to the BS and all femtocells are disabled, while Figure 4 illustrates the total energy consumption of the BS. The horizontal axis depicts the simulation time, the vertical axis on the left depicts the energy consumption of the BS and the vertical axis on the right depicts the number of terminals that are connected to the BS. As it can be observed, the number of terminals that are connected to the BS is constantly 50 due to the fact that there are no nearby femtocells to acquire terminals.

Figure 5 illustrates the progress of the transmission power of the BS. The horizontal axis depicts the simulation time and the vertical axis on the left depicts the transmission power of the BS. The black line depicts the moving average of the transmission power of the BS. As it can be observed, when

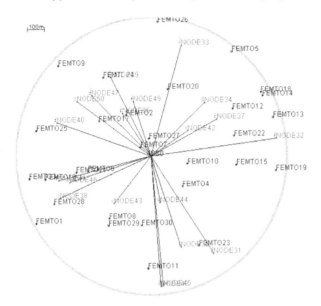

Figure 3 A macro BS without femtocells.

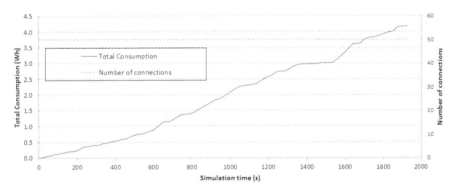

Figure 4 Total energy consumption of a BS without femtocells.

more terminals are transmitting data, the transmission power is high, while when most terminals are idle, the transmission power is very low.

Figure 6 depicts a snapshot from the case of the macro BS with 10 nearby femtocells, while Figure 7 illustrates the total energy consumption of the BS with 10 nearby femtocells. The horizontal axis depicts the simulation time, the vertical axis on the left depicts the energy consumption of the BS and the vertical axis on the right depicts the number of terminals that are connected

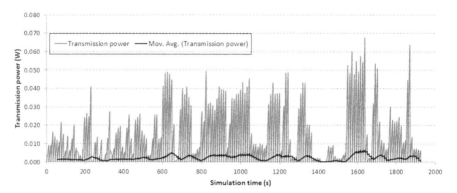

Figure 5 Transmission power of the BS without femtocells.

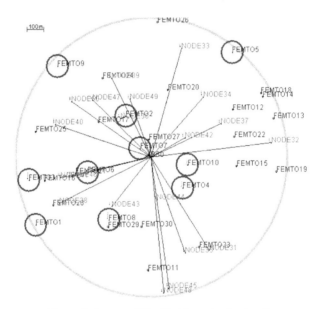

Figure 6 A macro BS with 10 nearby femtocells.

to the BS. Apparently, the energy consumption of the BS decreased due to the fact that the femtocells acquired a proportion of the traffic of the macro BS. In addition, as it can be observed the number of terminals that are connected with the BS has reduced since a proportion of the terminal is served by the femtocells.

Figure 8 illustrates the progress of the transmission power of the BS. The horizontal axis depicts the simulation time and the vertical axis on the left

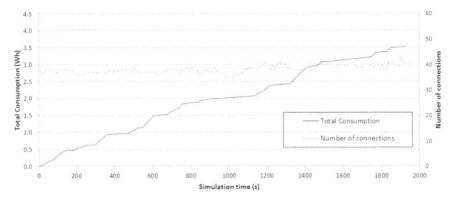

Figure 7 Total energy consumption of the BS with 10 nearby femtocells.

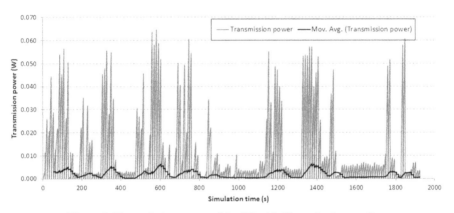

Figure 8 Transmission power of the BS with 10 nearby femtocells.

depicts the transmission power of the BS. The black line depicts the moving average of the transmission power of the BS.

Figure 9 depicts a snapshot from the case where 20 femtocells are within the area of the macro BS, while Figure 10 illustrates the total energy consumption of the BS. The horizontal axis depicts the simulation time, the vertical axis on the left depicts the energy consumption of the BS and the vertical axis on the right depicts the number of terminals that are connected to the BS. Apparently, the energy consumption of the BS has decreased even more due to the fact that more femtocells acquired traffic from the macro BS. In addition, as it can be observed the number of terminals that are connected with the BS has reduced more.

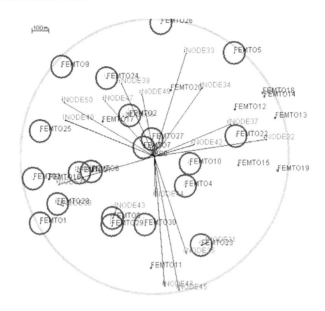

Figure 9 A macro BS with 20 nearby femtocells.

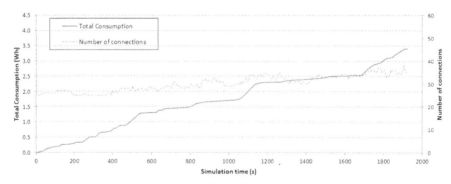

Figure 10 Total energy consumption of the BS with 20 nearby femtocells.

Figure 11 illustrates the progress of the transmission power of the BS. The horizontal axis depicts the simulation time and the vertical axis on the left depicts the transmission power of the BS. The black line depicts the moving average of the transmission power of the BS.

A snapshot of the case of a macro BS with 30 nearby femtocells is depicted in Figure 2. Figure 12 illustrates the total energy consumption of the BS with 30 nearby femtocells. The horizontal axis depicts the simulation time,

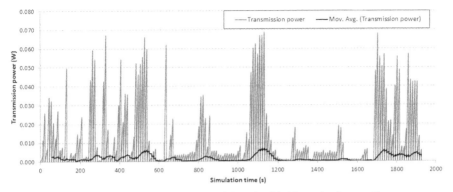

Figure 11 Transmission power of the BS with 20 nearby femtocells.

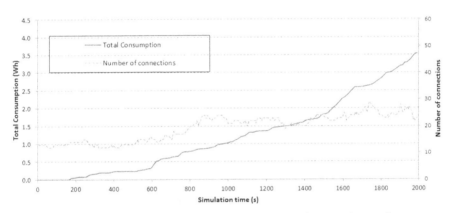

Figure 12 Total energy consumption of the BS with 30 nearby femtocells.

the vertical axis on the left depicts the energy consumption of the BS and the vertical axis on the right depicts the number of terminals that are connected to the BS. In this case, the energy consumption of the BS has not decreased that much in comparison with the case of the 20 nearby femtocells, since most users have been already rerouted to the femtocells.

Figure 13 illustrates the progress of the transmission power of the BS. The horizontal axis depicts the simulation time and the vertical axis on the left depicts the transmission power of the BS. The black line depicts the moving average of the transmission power of the BS.

Finally, Figure 14 illustrates a comparison of the energy consumption of the BS for the aforementioned cases. The vertical axis indicates the BS total energy consumption, while the horizontal axis depicts the network type,

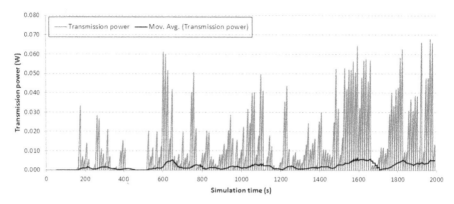

Figure 13 Transmission power of the BS with 30 nearby femtocells.

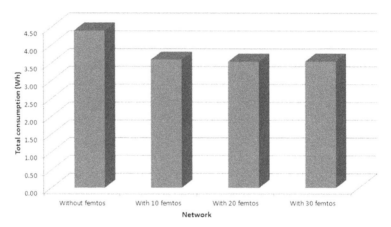

Figure 14 Energy consumption comparison for a BS with 0, 10, 20 and 30 nearby femtocells.

i.e. a network without femtocells and with 10, 20 and 30 femtocells. More specifically, when 10 femtocells were enabled to the network the energy consumption of the BS decreased by 18.54%. When 20 femtocells were enabled in the network, the energy consumption of the BS was reduced by 19.72% in comparison with the case were there were no femtocells. Finally, when 30 femtocells were enabled, the energy consumption of the BS was reduced by 20.01% percent. Therefore, as femtocells are deployed to the network, the energy consumption of the BS decreases. However, after a point where most of the traffic has been rerouted to the femtocells, the addition of more femtocells is not that much beneficial.

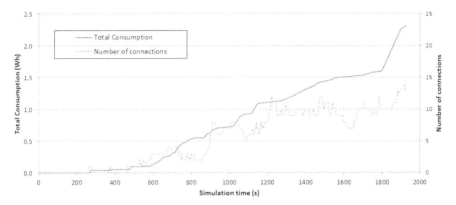

Figure 15 Total energy consumption of the BS with 30 terminals.

5.2.2 Test Case 2

In this test case, the impact of the number of terminals on the power consumption of the BS will be examined. A macro BS area with 30 femtocells is considered as in the previous case. The number of terminals for which the BS energy consumption will be studied is 30, 40 and 50 terminals.

Figure 15 illustrates the total energy consumption of the BS with 30 terminals. The horizontal axis depicts the simulation time, the vertical axis on the left depicts the energy consumption of the BS and the vertical axis on the right depicts the number of terminals that are connected to the BS. In this case, the energy consumption of the BS is very low since most users are served through the femtocells as can be observed also from the number of terminals that are connected to the BS.

Figure 16 illustrates the total energy consumption of the BS with 40 terminals. The horizontal axis depicts the simulation time, the vertical axis on the left depicts the energy consumption of the BS and the vertical axis on the right depicts the number of terminals that are connected to the BS. In this case, the energy consumption of the BS has increased as it was expected since more terminals were added to the network. Moreover, it can be observed that the number of the connected terminals to the BS has also increased.

The total energy consumption of the BS for the network that comprises 50 terminals is illustrated in Figure 12 of test case 1. As it can be seen, the energy consumption of the BS is even higher since 10 additional terminals were included in the network. In addition, the number of terminals that are connected with the BS was increased respectively.

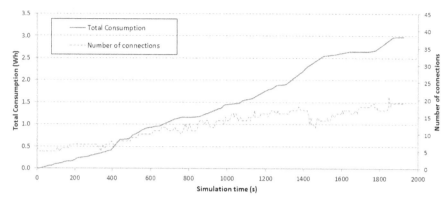

Figure 16 Total energy consumption of the BS with 40 terminals.

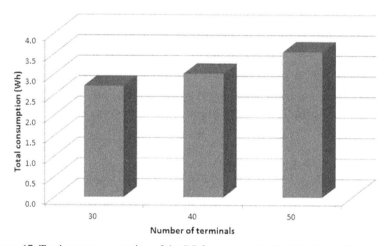

Figure 17 Total energy consuption of the BS for a network with 30, 40 and 50 terminals.

Finally, Figure 17 illustrates a comparison of the total energy consumption of the BS for the above-mentioned cases. More specifically, for the case of the network with 30 terminals, the consumption of the BS was 2.71 Wh. When 10 more terminals were added to the network, the BS consumption increased by 10% and reached 3.01 Wh. Finally, when terminals increased to 50, the BS consumption increased by 15% in comparison with the case of 40 terminals. As a result, as it was expected the energy consumption of the BS tends to increase when the number of terminals also increases.

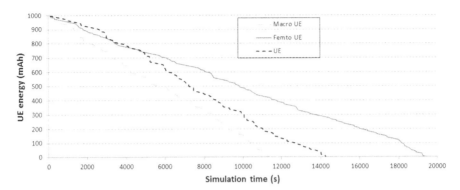

Figure 18 Progress of the energy level of macro-terminals, femto-terminals and terminals that connect both to macro BSs and femtocells.

5.2.3 Test Case 3

In this test case, the battery consumption of terminals will be examined. More specifically a terminal that is served exclusively through the macro BS (macro User Equipment (UE)), a terminal that is served only through a femtocell (femto UE) and a terminal that is moving and can be served from the BS and a femtocell (UE), but not simultaneously will be studied.

Figure 18 depicts the progress of the energy level of a macro UE vs. the progress of the energy level of a femto UE vs. the progress of the energy level of a UE. It is assumed that the energy of a terminal decreases as time progresses and that also decreases when data are transmitted. In addition, it is assumed that the terminals send data every 30 secs. Apparently, the energy of the femto UE decreases at a lower rate since the terminal needs less transmission power to communicate with the femtocell, while the energy of the macro-terminal decreases in a high rate due to the high transmission power. On the other hand, the progress of the energy level of the moving terminal is in the middle of the progress of the femto-UE and the macro-UE as it was expected since when it is within the coverage of an available femtocell, i.e. in terms of capacity and accessibility, it connects to the femtocell, otherwise it connects to the BS. More specifically, as it can be observed in Figure 19, the battery lifetime, i.e. the time needed for the energy level of a terminal to reach 0, of a femto-UE is 19506, while for a macro UE the lifetime is 11400 and for the moving UE is 14263. Hence, the lifetime of a macro-UE is 41.56% lower than the one of a femto-UE and 25.12% lower than the one of the moving UE.

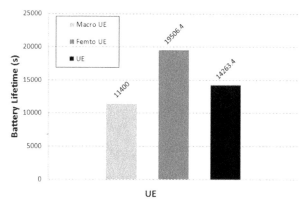

Figure 19 Battery lifetime of macro-terminals, femto-terminals and terminals that connect both to macro BSs and femtocells.

6 Conclusion

This paper considers ONs as a promising solution to the problem of capacity expansion of congested infrastructure. In particular, it focuses on the benefits that derive from the use of femtocells at the energy consumption of a macro BS and the battery lifetime of terminals through 3 indicative test cases. Simulation results depicted that the BS energy consumption is decreasing while the number of femtocells increases. The reason is that the more femtocells, the more traffic is acquired by the BS and therefore the number of terminals that connect to the BS is low.

In addition it was illustrated that the BS energy consumption increases while the number of terminals increases in the network. The reason is that the more terminals in the network, the more connections will be made with the BS. As a result the BS needs higher transmission power to serve the terminals.

Furthermore, the battery lifetime of a macro-terminal, a femto-terminal and a moving terminal was studied. Simulations illustrated that the battery lifetime of a femto-terminal is significantly higher than the one of the macro-terminal.

As a result, capacity extension through femtocells proved to be an energy efficient solution that provides benefits not only to the operator than is able to make savings in OPEX, but also to the end-user that can experience higher QoS levels and higher battery lifetime.

Acknowledgment

This work is partially performed in the framework of the European-Union FP7 funded project OneFIT (www.ict-onefit.eu) and the E3 project National Participation funded by the General Secretariat of Research and Technology (GSRT) of the Greek Ministry of Development. The views expressed in this document do not necessarily represent the views of the complete consortium. The Community is not liable for any use that may be made of the information contained herein.

References

[1] V. Stavroulaki, K. Tsagkaris, M. Logothetis, A. Georgakopoulos, P. Demestichas, J. Gebert, and M. Filo. Opportunistic networks: An approach for exploiting cognitive radio networking technologies in the Future Internet. *IEEE Vehicular Technology Magazine*, 52–59, September 2011.

[2] A. Georgakopoulos, K. Tsagkaris, V. Stavroulaki, and P. Demestichas. Efficient opportunistic network creation in the context of Future Internet. In J. Domingue et al. (Eds.), *The Future Internet: Achievements and Technological Promises*, Lecture Notes in Computer Science, Vol. 6656. Springer, pp. 293–306, 2011.

[3] B. Han, P. Hui, V. Anil Kumar, M. Marathe, J. Shao, and A. Srinivasan. Mobile data offloading through opportunistic communications and social participation. *IEEE Transactions on Mobile Computing*, 2011.

[4] S. Dimatteo, P. Hui, B. Han, and V. Li. Cellular traffic offloading through WiFi networks. In *Proceedings of 8th IEEE International Conference on Mobile Ad-hoc and Sensor Systems (IEEE MASS 2011)*, Valencia, Spain, October 2011.

[5] S. Yeh, S. Talwar, G. Wu, N. Himayat, and K. Johnsson. Capacity and coverage enhancement in heterogeneous networks. *IEEE Wireless Communications*, 18(3), June 2011.

[6] N. Himayat et al. Heterogeneous networking for future wireless broadband networks. IEEE C80216-10_0003r1, January 2010.

[7] A. Agustin, J. Vidal, and O. Muñoz. Interference pricing for self-organisation in OFDMA femtocell networks. In *Proceedings of European Workshop on Broadband Femtocell Networks, Future Network and Mobile Summit*, Warsaw, Poland, pp. 1–8, June 2011.

[8] S. Barbarossa, A. Carfagna, S. Sardellitti, M. Omilipo, and L. Pescosolido. Optimal radio access in femtocell networks based on markov modeling of interferers' activity. In *Proceedings of IEEE International Conference on Acoustics, Speech and Signal Processing (ICASSP)*, Prague, pp. 3212–3215, May 2011.

[9] P. di Lorenzo and S. Barbarossa. Distributed resource allocation in cognitive radio systems based on social foraging swarms. In *Proceedings of IEEE 11th International Workshop on Signal Processing Advances in Wireless Communications (SPAWC)*, June 2010.

[10] C. Abgrall, E.C. Strinati, and J.C. Belfiore. Distributed power allocation for interference limited networks. In *Proceedings of IEEE 21st International Symposium on Personal In-

door and Mobile Radio Communications (PIMRC), Istanbul, pp. 1342–1347, September 2010.

[11] P. Mach and Z. Becvar. QoS-guaranteed power control mechanism based on the frame utilization for femtocells. *EURASIP Journal on Wireless Communications and Networking*, 2011.

[12] P. Skillermark, M. Olsson, Y. Jading, G. Auer, V. Giannini, M.J. Gonzalez, and C. Desset. System level energy efficiency analysis in cellular networks. In *Proceedings of Future Network & Mobile Summit 2011*, Warsaw, Poland. IIMC International Information Management Corporation, June 2011.

[13] B. Debaillie, A. Giry, M.J. Gonzalez, L. Dussopt, M. Li, D. Ferling, and V. Giannini. Opportunities for energy savings in pico/femto-cell base-stations. In *Proceedings of Future Network & Mobile Summit 2011*, Warsaw, Poland. IIMC International Information Management Corporation, June 2011.

[14] ThinkFemtocell, available online at http://www.thinkfemtocell.com/.

[15] G. de la Roche, A. Valcarce, D. Lopez-Perez, and J. Zhang. Access control mechanisms for femtocells. *IEEE Communications Magazine*, 48(1):33–39, January 2010.

[16] M. Hata. Empirical formula for propagation loss in land mobile radio services. *IEEE Transactions on Vehicular Technology*, VT-29(3): 317–325, August 1980.

[17] Z. Jako and G. Jeney. Downlink femtocell interference in WCDMA networks. In *Proceedings of Energy-Aware Communications, 17th International Workshop (EUNICE 2011)*, Springer, September 2011.

[18] A. Keranen, J. Ott, and T. Karkkainen. The ONE simulator for DTN protocol evaluation. In *Proceedings of SIMUTools'09: 2nd International Conference on Simulation Tools and Techniques*, Rome, March 2009.

Biographies

Dimitrios Karvounas conducts research at the Department of Digital Systems in University of Piraeus. His main interests are cognitive and autonomous management techniques of the Future Internet and bio-inspired optimization algorithms.

Andreas Georgakopoulos conducts research at the Department of Digital Systems in University of Piraeus. His main interests include the specification of cognitive management mechanisms for the creation and maintenance of dynamic access networking paradigms in the FI era, such as opportunistic networking, etc.

Dimitra Panagiotou is a post-graduate student at the Department of Digital Systems in University of Piraeus. She conducts research in the context of techno-economic management of digital systems.

Vera Stavroulaki is an Assistant Professor at the Department of Digital Systems in University of Piraeus. Her main interests include cognitive management functionality for autonomous, reconfigurable user devices (operating in heterogeneous wireless networks), service governance and virtualization.

Kostas Tsagkaris is a senior research engineer and adjunct lecturer at the Department of Digital Systems in University of Piraeus. His current interests are in the design and management of autonomic/cognitive networks, optimization algorithms, and learning techniques.

Panagiotis Demestichas leads the laboratory of Telecommunication Networks and integrated Services (TNS) of the University of Piraeus. His research interests include the design, management and performance evaluation of mobile and broadband networks, algorithms and queuing theory.

Survey on Energy Efficiency in Office and Residential Computing Environments

Andreas Berl[1], Gergő Lovász[1], Hermann de Meer[1] and
Thomas Zettler[2]

[1]*Computer Networks and Computer Communications, University of Passau,
94032 Passau, Germany; e-mail: berl@uni-passau.de*
[2]*Lantiq GmbH, Munich, Germany*

Received 2 February 2012; Accepted: 10 April 2012

Abstract

Energy efficiency of computing equipment in office and residential environments gets more and more important, with respect to the world-wide desire to reduce CO_2 emissions and the increasing cost of energy. While hardware itself gets cheaper, the cost of energy begins to dominate the total cost of ownership of a product. This paper gives an overview on energy saving methods that are applied today, with a special focus on office and residential environments. Currently used methods are classified into three categories: (1) autonomous management of devices, (2) coordinated management of devices, and (3) coordinated management of services. Various implementations of these methods in office and residential environments are described and compared to each other. The comparison illustrates possible directions of future research in the area of energy efficiency.

Keywords: energy efficiency, office environments, residential environments.

Journal of Green Engineering, Vol. 2, 255–272.

1 Introduction

Energy-efficient Information and Communication Technology (ICT) is fostered by labels such as the U.S. Energy Star [8] or the European TCO Certification [16, 22, 35]. Furthermore, regulations as the European Energy related Products Directive [4] and the European Codes of Conduct [36] rate IT equipment according to their environmental impact.

The strict separation between the often used terms energy efficiency for ICT (making ICT energy efficient) and ICT for energy efficiency (using ICT to achieve energy efficiency) [52] is vanishing in the area of residential and office environments. More and more non-IT equipment joins the network (i.e., gets an IP-address and becomes manageable [42, 44]), as the Internet of Things embraces more and more devices. This opens up the opportunity to save energy in classical non-IT equipment as well as in IT equipment by using the same management mechanisms. The U.S. Energy Information Agency [17] reports, that home electronics including IT equipment as PCs and entertainment TV sets account for 7% of the electricity consumed by U.S. households. Moreover, the European Eco-Design Directive [4] shows in recent studies [31] that IT equipment as PCs, peripherals, printers or phones exhibits in total consumes more energy than data centers. The carbon footprint that is related to usage and directly corresponds to the energy consumption is shown to be 259 Mt CO_2 in 2002 and predicted to be 640 Mt CO_2 in 2020 (60 and 59% share of the global ICT footprint). For example the Telecom device's global footprint was 18 Mt CO_2 in 2002 and is expected to increase almost threefold to 51 Mt CO_2 by 2020 driven mainly by rises in the use of broadband modems/routers and IPTV boxes.

Obviously the energy-saving potential in residential and office computing environments is huge, but due to their distributed nature and heterogeneous device landscape hard to exploit. This paper analyses energy-saving methods that are available for IT equipment. It classifies current work into three main categories of energy-saving methods: (1) Autonomous management of devices enables the reduction of energy consumption locally at a single device (e.g., by built-in energy-efficiency features). (2) Coordinated management of devices enables the optimization of the energy consumption of a group of devices that actively exchange energy-related information. (3) Coordinated management of services enables the replacement, delegation, and consolidation of services and aims at optimizing the energy consumption of a service or a group of services. In addition, this paper explores and compares implementations of energy-saving methods of each category in the context of office and residential environments.

The remainder of this paper is structured as follows: Section 2 categorizes current energy-saving methods. Sections 3 and 4 analyse the application of these methods in office and residential environments. Section 5 provides a comparison of energy-saving methods concerning their use in both environments, and Section 6 concludes this paper.

2 Energy-Saving Methods

This section identifies three disjunctive categories of energy-saving methods that reduce the energy consumption of devices. For each category several examples are described.

2.1 Autonomous Management of Devices

Autonomous management of devices covers energy management methods that reduce the energy consumption of a device without coordination with other devices or the user. Instead, the device exploits its built-in energy efficiency features autonomously. *Dynamic external condition adaption* monitors conditions that are caused externally (as CPU-workloads, CPU-temperatures, or user-interaction) and manages parts of a device accordingly.

The goal of the adaption is to dynamically adapt the managed device to its environment in a way that the energy consumption of the device is reduced. It is important to see that this happens without a conscious interaction of the user. Examples of dynamic external condition adaption are:

- A monitor is dimmed in reaction to low light conditions.
- A fan is slowed down if the CPU is below a certain temperature.
- Hardware parts are incrementally turned off due to sensing a lack of user-machine interaction (e.g., display or disk).

2.2 Coordinated Management of Devices

In contrast to the autonomous management of devices, the coordinated management of devices addresses the cooperation between devices.

Automatic coordination reduces the energy consumption of a set of devices by exchanging energy-related information that eases up energy management decisions. The purpose is to reduce the energy consumption of a whole set of devices instead of locally optimizing the energy consumption for each single device. Inter-device coordination can be achieved in a centralized or decentralized way. In a centralized coordination approach, a centralized entity either polls information from the managed devices or the managed devices inform the managing entity periodically or at the occurrence of an energy-relevant event. Based on the gathered information and policies, the central entity instructs devices to apply power saving methods. Coordination can also be achieved in a decentralized way, where energy-related information is exchanged, but decisions are made based on the local view of each device. *User-based coordination* is triggered by implicit or explicit interaction between user and device. On one hand, the device may push information to the user, e.g., a visualization of the current energy consumption of the device. On the other hand, the user is able to directly control the device, e.g., by sending the device to hibernation mode actively. Besides the energy savings that are directly achieved by this approach, there are additional psychological effects that foster the energy-efficient behaviour of a user: Immediate feedback on the effects of his actions motivates energy-efficient behaviour. Also competitive situations between users may be established, further motivating users to behave energy efficiently.

Examples of coordinated management of devices are:

- Cisco's EnergyWise [33] (see Section 3.2) represents a centralized management approach. A centralized server powers up/down groups devices, e.g., according to working/non-working times.
- The Energy Efficient Ethernet (EEE) standard (IEEE 802.3az [5]) is an example of decentralized management approach. During times without demand of data transmission, devices negotiate a low-power idle mode.
- Products as Kill-A-Watt [45] or Watts Up [19] (see Section 3.2) are products that support user-based coordination. They adapt their energy consumption to user behaviour and visualize consumed energy.
- Projects that have been performed in residential environments [43, 51] have shown that real-time feedback on power consumption leads to a reduction of energy consumption by up to 10%.

2.3 Coordinated Management of Services

Although services (e.g., print-servers, Open VPN servers, peer-to-peer clients, or user desktops) do not consume energy directly, they utilise devices and cause energy consumption indirectly. This category of methods reduces the energy consumption of services, by replacing, delegating, or consolidating them.

Service replacement is an approach for energy saving were services are replaced by more energy-efficient services that provide the same (or similar) functionality. Although the energy-saving effect of service replacement can be large, the overall impact on energy consumption is difficult to assess. It may even happen that the overall energy consumption increases, if the so-called rebound effect occurs. This effect describes the situation that a new energy efficient service is so attractive to users that a high demand is created which partially or fully compensates for the energy-saving effect of the replacement. *Service delegation* allows the transfer of a service from one device to another, e.g., from a non-energy efficient to an energy-efficient device or to an always-on device (e.g., a router). The main goal of service delegation is to allow under-utilised devices to delegate their services to other devices and change to an energy-saving mode. *Service consolidation* is based on the ability of devices to process more than a single service at the same time. The goal of service consolidation is to reorganise the service to device mapping within a group of devices in order to minimise the number of utilised devices. This means that the utilisation of some devices is increased while other devices are relieved from their duties. Unutilised devices are hibernated to save energy. Service consolidation can be done statically or dynamically. If it is done statically, a set of devices is determined that processes all required services. If the external circumstances change, the allocation of the services is not dynamically adapted. Dynamic service consolidation, in contrast, allows for the relocating of devices when external circumstances change, e.g., the loads of services change, or a device fails. Examples of service replacement, delegation, and consolidation are:

- Terminal servers [13,14,20] and virtual desktop infrastructures [3,18,21] (see Section 3.3) replace user desktops in office environments. Instead the desktops are consolidated on servers within the data centre.
- Virtual private network server services can be delegated to the home gateway (router) in residential environments.
- Cloud computing achieves energy efficiency [28, 37, 48] (see Section 4.3) by consolidating user services (e.g., storage services) within data centres. Cloud providers achieve a high utilisation of hardware and customers can dynamically allocate and release resources in the cloud.

3 Office Environments

Office computing environments consist, e.g., of office hosts, network, peripheral devices as monitors, printers, scanners, and IP-phones. Within office environments, especially office hosts contribute significantly to the IT related energy consumption. On one hand, there is a high number of such hosts because usually each employee typically has his own host. On the other hand, office hosts are often turned on 24/7. Webber et al. [50] have analyzed sixteen office sites in the U.S. and reported that 64% of all investigated office hosts were running during the nights. Although such hosts are mostly idle (CPU usage of 0%) during the time they are turned on, it is important to see that they still consume a considerable amount of energy. Measurements that have been performed at the University of Sheffield [32] show that typical office hosts which are idle still consume 49 to 78% of the energy that they need when they are intensely used, leading to an immense waste of energy.

3.1 Autonomous Management of Devices

Current office computing equipment often has the ability of saving energy by falling into low-power states if it remains unused for a critical period of time. Hosts, monitors, or printers are dynamically hibernated to save energy. The low-power states of office hosts can be configured by the user and kick in when a host is idle for a critical time period. The Advanced Configuration and Power Interface (ACPI) specification [38] defines four different power states that an ACPI-compliant computer system (e.g., an office host) can be in. These states range from G0-Working to G3-Mechanical-Off. The states G1 and G2 are subdivided into further sub-states that describe which components are switched off in the particular state. Separate power states (D0-D3 for sub-devices and C0-C3 for CPUs) are defined, similar to the global power states [7, 34, 38]. However, as a matter of fact, many devices that are low-power capable do not successfully enter these states. Low-power modes are subject to the complex combined effects of hardware, operating systems, drivers, applications – and after all – the user-based power management configuration. Webber et al. [50] report that in the investigated offices only 4% of all hosts actually have switched to low-power modes during the night.

3.2 Coordinated Management of Devices

In office environments power management solutions are able to optimise the energy consumption of hosts that remain turned on while their users are absent. Examples of this approach are eiPowerSaver [6], Adaptiva Companion [2], FaronicsCore [9], KBOX [11], or LANrev [1]. office-wide power management policies are applied in such approaches. Office hosts are forced to adopt power management configurations, independent of user settings. Therefore, idle hosts can be set to a low-power state or be powered off to save energy. Additionally, often mechanisms are provided by such approaches to wake up hosts if necessary (e.g., based on Wake-on-LAN technology). This way, inactive hosts can be accessed for administrative jobs (e.g., backups that happen during the night) and for remote usage. Cisco's EnergyWise controls office equipment that is powered by Power over Ethernet (PoE). It can be used to power down IP-phones during nights and to power them up again in the mornings. Additionally, EnergyWise can be used to apply energy management to hosts and to report energy savings within the office.

There are also user-based coordination methods available in offices: The approach of Greentrac [12] is setting its focus on the user's energy awareness. A user is periodically informed about the energy consumption of the devices he is using. If the user is aware of the energy consumption he causes, he is able to change his behaviour in order to save energy. The Greentrac-approach uses incentives to motivate the employees to implement energy-saving measures.

3.3 Coordinated Management of Services

A typical example of service replacement and consolidation within office environments, is the replacement of energy-consuming office hosts by highly energy-efficient *thin clients* [49]. *Terminal-server* approaches, e.g., move user desktops to centralized terminal servers that are able to serve multiple users simultaneously (consolidation). Terminal-server solutions are based on multi-user concepts where several users are able to log-on to a single OS that is provided by the terminal server. OS, applications, and user data are stored in the data centre and can be remotely accessed by thin clients. Common terminal server software products are Citrix XenApp [20], Microsoft Windows Server 2008 [14], or the Linux Terminal Server Project [13].

In the *Virtual Desktop Infrastructure (VDI)* approach each user gets his own Virtual Machine (VM). Similar to terminal servers, the VMs are stored within the data centre and can be accessed remotely by energy-efficient thin clients or any host with remote desktop software. In contrast to the terminal server approach, the VDI approach has the advantage that each user can utilise his preferred OS and individual applications (not all standard applications are able to run on terminal servers) and new virtualised desktops can be

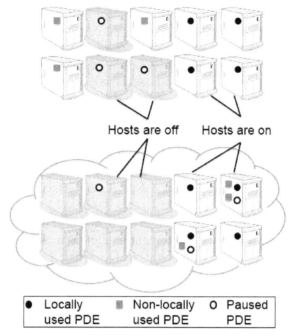

Figure 1 Common and virtualised office.

easily deployed. Furthermore, the virtualised desktops are strictly isolated from each other, while being managed within the data centre. However, it is important to see that the provision desktop environments within VMs is demanding: All of the VMs need a sufficient amount of CPU cycles, RAM, disc I/O, and other hardware resources to operate. Therefore, the number of VMs that can be provided by a single server is rather limited. VDI products are, e.g., VMWare View [3], Citrix XenDesktop [21], or Parallels Virtual Desktop Infrastructure [18].

In [24–26,30], a virtualised office environment is suggested that achieves a consolidation of services within office environments, independent of data centre equipment. Office hosts are virtualised and virtual desktops are consolidated dynamically on office hosts. Whereas terminal server and VDI solutions impose changes to the office environment (thin clients replace full featured office hosts), the virtualised office environment utilises available office hosts. Terminal servers and VDIs, instead, move office services into the data centre, which has two main disadvantages: First, additional hardware needs to be purchased and managed, second, the additional data centre hardware consumes energy itself. Data centre equipment typically consumes more energy than desktop hosts [41], due to high-performance parts, parts that provide redundancy, and, especially, the cooling that needs to be applied within the data centre.

In Figure 1 the transition from an ordinary to a virtualised office environment is illustrated. It can be observed in the upper part of the figure that in the ordinary office environment the Personal Desktop Environments (PDEs) and the hosts are interdependent. Seven hosts are turned on together with seven PDEs and three hosts (with PDEs) are turned off. The situation is different in the virtualised office environment shown in the lower part of the figure. Although the number of currently running PDEs is the same as before, only four hosts are actually turned on. It can be observed, e.g., that the upper right host is providing three PDEs to users simultaneously.

Possible savings of about 50% of energy are reported with this approach [24]. In comparison to VDI solutions which are able to save energy for office environments with more than 25 hosts [40], the virtualised office environments saves already energy in offices with only 4 hosts.

4 Residential Environments

More and more end-users have residential computing networks that consist of desktops, laptops, game pads, or home theater PCs. Such devices provide all kind of applications, including client, server, or peer-to-peer. Typically, residential networks have a gateway (residential gateway) to the Internet, e.g., a Digital Subscriber Line (DSL) router.

Although office and residential environments seem to be similar on a first glance, there are some major differences: in the office environment rather homogeneous office hosts are interconnected via Fast/Gigabit Ethernet and administered by a professional administrator. In residential environments, highly heterogeneous hosts are usually connected to the Internet via DSL-connections, which typically have asynchronous up/download capacities and are administered by individual users. The access technology can be both, wired as well as wireless. In residential environments, the users may be children or adults with certain rights and restrictions in the home network. This prevents the use of uniform security policies.

4.1 Autonomous Management of Devices

Similar to hardware of office environments also residential equipment performs a dynamic adaption to external conditions. Home IT-equipment (as laptops or printers) is able to sense a lack of user interaction in order to turn into a hibernation mode. The residential gateway is a device that is constantly turned on and typically provides a wide range of autonomous power management features. Even in the case of no user interaction, a large number of functions have to remain active to guarantee good user experience and to fulfill industry standards:

- A DSL-IP connection is required to receive VoIP calls. To transmit the IP stream the physical layer has to be kept active.
- The WLAN base station has to transmit beacon signals in order to perform the association of new mobile devices to the WLAN network and to maintain the wireless link to previously associated devices.
- Ethernet link detection has to be active and attached devices have to be managed when requesting a new link.
- For the DECT/CatIQ cordless telephony interface the incoming call detection has to be assured for all interfaces.
- The attachment of new devices to USB needs to be detected.

The list indicates some minimum active functions which have to be maintained during autonomous management. In addition, other services may be required to like for example FTP server functionality, multimedia server or home automation functions. If the user actively decides not to use WLAN during night time,e.g, it can be turned off by using an autonomous timer based management that is configurable by the user.

4.2 Coordinated Management of Devices

Home automation, e.g., provides a coordinated management of non-IT devices in the residential environment. Standards as G.hn [23] allow the connection of devices over any wire (power line, coax cable, phone line) or Wi-fi wireless connection.

Studies of the U.S. Energy Information Agency [17] show that major energy consumption in households stems from classical non-IT equipment. Therefore home automation opens a promising new opportunity for power saving by including additional information for example from sensor networks to energy management decisions.

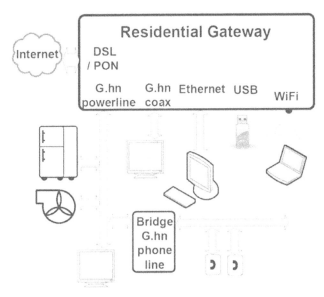

Figure 2 Home automation network.

An example of a home automation network architecture is illustrated in Figure 2, where a residential gateway is connecting multiple physical media forming together a heterogeneous network reaching virtually every controllable device. A DSL or Passive Optical Network (PON) interface offers Wide Area Network (WAN) services using broadband residential network. G.hn standard ports are used for coaxial cable and powerline communication. Phones and phone-line devices are included via an inter-domain bridge. Ethernet LAN and USB devices are also covered. Residential gateways allow the processing of automation applications inside the gateway independent from running PCs, which substantially saves power. Residential gateways are typically "always on" devices and form a natural central point for home networking.

Also user-based coordination is achieved within the residential environment. Emerging technologies as, e.g., smart metering approaches raise the user's energy consumption awareness. Energy consumption can be monitored locally and remotely [47]. This is important to keep users aware about their energy behaviour. Having real-time feedback on current energy consumption allows the user to link his actions to an increase of energy consumption. Taherian et al. [46], e.g., describe an energy monitoring system for resid-

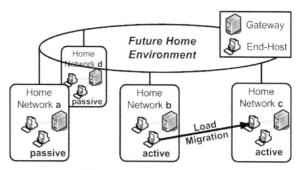

Figure 3 FHE architecture.

ential and office environments that supports continuous real-time feedback on energy consumption.

4.3 Coordinated Management of Services

Also service delegation and consolidation is achieved in residential environments. In this field the residential gateway plays a key role. Residential gateways mediate access to the Internet, run services on the user's behalf, and control IT and non-IT equipment. Furthermore, residential gateways are able to support the coordinated management of services and are able to provide services themselves energy efficiently: Residential gateways can provide access to peripheral devices as USB-Disks, printers, Network Attached Storage (NAS), multiport routing, switching and encryption, VoIP telephony functions (e.g., DECT/CatIQ or VoIP FXS/FXO H.323), or they can run P2P software as BitTorrent or eDonkey.

Based on such residential gateways, Berl and co-authors [27–29, 39] describe a Future Home Environment (FHE) that enables an energy-efficient consolidation of services in home networks. It suggests the sharing of home network resources with users of other home networks, similar to Grid computing approaches. Load is shifted to a small number of computers, in order to relieve others. Unloaded computers are hibernated or turned off. A home network is called active if it contains at least one computer which is turned on and can share resources. In a passive home network only the gateway is on-line and other hardware is hibernated or turned off.

The FHE architecture is illustrated in Figure 3. In this example four home networks are interconnected by the FHE overlay, two active and two passive homes. In the figure load is migrated from an end-host in the active home network (b) to an end-host in the active home network (c). The end-host in home network (b) can be hibernated or turned off after the migration process. If no further computer is turned on in home network (b), it can change its status to passive.

5 Comparison

Most of the energy-saving methods that have been described in Section 2 are applied in office and residential computing environments (see Sections 3 and 4). Whereas the autonomous management of devices is implemented similarly in office and residential environments (hibernation of unused devices), the coordinated management of devices is implemented diversely in the two environments. On the one hand, office-wide power management approaches and the controlling of PoE devices is applied in office environments. Such mechanisms can be easily applied, due to the rather homogeneous office computing environment and similar usage patters of office users, which eases up the energy management: Sets of devices can be hibernated, e.g., according to time-based energy-saving policies (working/non working times). Multipurpose devices of residential networks, on the other hand, are rather heterogeneous and the behaviour of the users is less predictable. The management of the devices needs to be done in a context-aware way, where the behaviour of the users is monitored in order to take management decisions. Also waking up devices is often easier in office environments as Wake-on-LAN can be applied to devices that are connected to wired networks. In residential environments many devices are attached wireless to the home gateway which makes it hard to wake them up remotely. Instead, home automation systems are applied that mainly adapt energy consumption of non-IT devices to the behaviour of persons in a household.

The user-device interaction for energy-saving requires users sufficiently trained and with awareness of energy consumption. The typical residential-user will need a simple and easy to use interface that provides direct feedback on energy consumption. Independent of residential or office environment the general consciousness of energy saving is key to motivate user to take action.

The application of coordinated management of services, however, is unbalanced between office and residential computing environments: Service delegation is typically applied in residential environments. Home Gateways take over services as, e.g., printing servers, network storage servers, or servers for DECT phones and needs to be further exploited in office environments. Together with the novel approach of the Virtualized Office Environment (as described in Section 3.3) the energy consumption of offices can be reduced, especially if no data centre infrastructure is available. Whereas service replacement and consolidation has already been applied widely in the area of office environments (in terms of virtual desktop infrastructures and terminal servers), it is not yet sufficiently exploited in residential environments. Novel approaches, as the Future Home Environment (see Section 4.3), provide a high potential of energy saving in this area. Also the paradigm of cloud computing can be further exploited: Instead of thin-clients, users of residential networks are able to use energy-efficient equipment as smart-phones, tablets, or netbooks to access cloud-based services. Gaming PCs, desktops, or home theater PCs may not be needed anymore in future residential network scenarios. There are already some approaches available, as e.g. OnLive [15] or GAIKAI [10], where demanding 3D games can be played within the cloud.

6 Conclusion

This paper has reviewed the state of the art of available energy-saving methods, especially concerning office and residential environments. Currently applied methods have been categorized into three classes: (1) The autonomous management of devices allow devices to reduce their energy consumption individually, without active cooperation of other devices or humans. (2) Coordinated management of devices covers automatic management as well as user interaction. In contrast to the first category, such strategies save energy through cooperation, using energy-relevant signalling to exchange information. Groups of devices are managed cooperatively (or are managing themselves) to achieve the common goal of a reduced energy consumption. Finally, (3) the coordinated management of services performs the replacement, delegation, and consolidation of services in a cooperative way to reduce the overall energy consumption.

Although all of the mentioned categories are applied within office environments as well as in residential environments, their application is unbalanced. Especially, the coordinated management of services provides a high potential of energy savings that can be exploited by future developments.

Acknowledgements

The research leading to these results has been partly supported by the German Federal Government BMBF in the context of the G-Lab_Ener-G project, by the ECs FP7 All4green project (grant agreement no. 288674), by the FP7 EuroNF Network of Excellence (grant agreement no. 216366, Joint Specific Research Project EEWMI) and by the COST Action IC0804.

References

[1] Absolute MANAGE. http://www.lanrev.com/solutions/power_management.html.

[2] Adaptiva Companion. http://www.adaptiva.com/products_companion.html.

[3] Desktop Virtualization: VMware View. http://www.vmware.com/products/view.

[4] Directive 2009/125/EC. http://eur-lex.europa.eu/LexUriServ/LexUriServ.do?uri=OJ:L: 2009:285:0010:0035:en:PDF.

[5] Draft Amendment to IEEE Std 802.3-2008,IEEE Draft P802.3az/D2.0, IEEE 802.3az Energy Efficient Ethernet Task Force. http://grouper.ieee.org/groups/802/3/az/ public/index.html.

[6] eiPower Saver Solution. http://entisp.com/pages/eiPowerSaver.php.

[7] Energy Star for External Power Adapters. http://www.energystar.gov/ia/partners/ product_specs/program_reqs/eps_prog_req.pdf.

[8] EU Energy Star: European Commission Directorate-General for Energy. http://www.eu-energystar.org.

[9] FARONICS: Intelligent Solutions for ABSOLUTE Control. http://faronics.com/html/CoreConsole.asp.

[10] GAIKAI Open Cloud Gaming Platform. http://www.gaikai.com.

[11] Green IT. http://www.kace.com/solutions/green-it.php.

[12] Greentrac – A new way to reduce the power your company uses by leveraging your greatest resource: Your people. http://www.greentrac.com/index.php.

[13] LTSP: LINUX Terminal Server Project. http://www.ltsp.org.

[14] Microsoft Windows Server 2008 R2. http://www.microsoft.com/windowsserver2008.

[15] OnLive desktop. http://www.onlive.com.

[16] TCO Certification. http://www.tcodevelopment.com.

[17] U.S. Energy Information Agency, Household Electricity Report, U.S. Department of Energy. http://www.eia.doe.gov/emeu/reps/enduse/er01_us_tab1.html.

[18] Virtuelle Desktop-Infrastruktur. http://www.parallels.com/solutions/vdi.

[19] watts up? https://wattsupmeters.com.

[20] XenApp. http://www.citrix.com/XenApp.

[21] XenDesktop 5: The virtual desktop revolution is here ... for everyone. http://www.citrix.com/virtualization/desktop/xendesktop.html.

[22] Preparatory studies for Eco-design Requirements of EuPs Lot 3 Personal Computers (desktops and laptops) and Computer Monitors. http://www.ebpg.bam.de/de/ ebpg_medien/003_studyf_07-08_complete.pdf, 2007.

[23] International Telecommunication Union, New ITU standard opens doors for unified 'smart home' network. http://www.itu.int/newsroom/press_releases/2009/46.html, October 2009.

[24] Andreas Berl. Energy Efficiency in Office Computing Environments. PhD Thesis, University of Passau, Fakultät für Informatik und Mathematik, May 2011.

[25] Andreas Berl and Hermann De Meer. A virtualized energy-efficient office environment. In *Proceedings of the ACM SIGCOMM 1st Int'l Conf. On Energy-Efficient Computing and Networking (e-Energy 2010)*, pages 11–20. ACM, April 2010.

[26] Andreas Berl and Hermann De Meer. An energy-consumption model for energy-efficient office environments. *FGCS*, 27(8):1047–1055, October 2011.

[27] Andreas Berl, Hermann De Meer, Helmut Hlavacs, and Thomas Treutner. Virtualization in Energy-Efficient Future Home Environments. *IEEE Communications Magazine*, 47(12):62–67, December 2009.

[28] Andreas Berl, Erol Gelenbe, Marco Di Girolamo, Giovanni Giuliani, Hermann De Meer, Minh Quan Dang, and Kostas Pentikousis. Energy-efficient cloud computing. *The Computer Journal*, 1–7, August 2009.

[29] Andreas Berl, Helmut Hlavacs, Roman Weidlich, Michael Schrank, and Hermann De Meer. Network virtualization in future home environments. In *Proc. of the 20th Int'l Workshop on Distributed Systems: Operations and Management (DSOM09)*, Lecture Notes in Computer Science (LNCS), Vol. 5841, pages 177–190. Springer Verlag, October 2009.

[30] Andreas Berl, Nicholas Race, Johnathan Ishmael, and Hermann De Meer. Network virtualization in energy-efficient office environments. *Computer Networks*, 54(16):2856–2868, November 2010.

[31] Mic Bowman, Saumya K. Debray, and Larry L. Peterson. SMART 2020: Enabling the low carbon economy in the information age. *ACM Trans. Program. Lang. Syst.*, 15(5):795–825, November 1993.

[32] Chris Cartledge. Sheffield ICT Footprint Commentary. http://www.susteit.org.uk/files/files/26-Sheffield_ICT_Footprint_Commentary_Final_8.doc, 2008.

[33] CISCO. Cisco EnergyWise Technology. http://www.tcodevelopment.com.

[34] European Commission. EU Directive for external power supplies. *Official Journal of the European Union*, April 2009.

[35] European Commission – DG TREN. CONNECT: Coordination and Stiumulation of innovative ITS activities in Central and Eastern European countries. http://www.connect-project.org/index.php?id=35.

[36] European Commission, Joint Research Centre. Code of conduct on energy consumption of broadband equipment, February 2011.

[37] Carl Hewitt. Orgs for scalable, robust, privacy-friendly client cloud computing. *IEEE Internet Computing*, 12(5):96–99, 2008.

[38] Hewlett-Packard. Microsoft, Phoenix, and Toshiba. Advanced configuration and power interface specification. *ACPI Specification Document*, 3, 2004.

[39] Helmut Hlavacs, Karin A. Hummel, Roman Weidlich, Amine M. Houyou, Andreas Berl, and Hermann De Meer. Distributed energy efficiency in future home environments. *Annals of Telecommunication: Next Generation Network and Service Management*, 63(9):473–485, October 2008.

[40] IGEL Technology. Große IT-Strategie für kleine Betriebe: VDI und Thin Clients. Press release, 30.3.2010.

[41] J.G. Koomey. Estimating total power consumption by servers in the US and the world, February 2007.

[42] Jörg Luther. *Living and Working in a Global Network*. Springer, Berlin/Heidelberg, 2005.

[43] L.T. McCalley and C.J.H. Midden. Energy conservation through product-integrated feedback: The roles of goal-setting and social orientation. *Journal of Economic Psychology*, 23(5):589–603, 2002.

[44] Steve Meloan. Toward a Global Internet of Things. Technical Report, Sun Developer Network, November 2003.

[45] P3 International. Welcome to P3 International. http://www.p3international.com.

[46] G. Coulouris, S. Taherian, M. Pias, and J. Crowcroft. Profiling energy use in households and office spaces. In *Proceedings of 1st International Conference on Energy-Efficient Computing and Networking 2010*, pages 21–30. ACM, New York, USA, 2010.

[47] H. Siderius and A. Dijkstra. Smart metering for households: Costs and benefits for the Netherlands. In *Energy Efficiency in Domestic Appliances and Lighting*, p. 207, 2006.

[48] L.M. Vaquero, L. Rodero-Merino, J. Caceres, and M. Lindner. A break in the clouds: Towards a cloud definition. *ACM SIGCOMM Computer Communication Review*, 39(1):50–55, 2008.

[49] Willem Vereecken, Lien Deboosere, Pieter Simoens, Brecht Vermeulen, Didier Colle, Chris Develder, Mario Pickavet, Bart Dhoedt, and Piet Demeester. Power efficiency of thin clients. *European Transactions on Telecommunications*, 13:1–13, 2009.

[50] C.A. Webber, J.A. Roberson, M.C. McWhinney, R.E. Brown, M.J. Pinckard, and J.F. Busch. After-hours power status of office equipment in the USA. *Energy – The International Journal*, 31(14):2487–2502, 2006.

[51] G. Wood and M. Newborough. Dynamic energy-consumption indicators for domestic appliances. *Energy and Buildings*, 35(8):821–841, 2003.

[52] Mazin Yousif. Towards green ICT. In *ERCIM News online edition*. European Research Consortium for Informatics and Mathematics (RCIM), October 2009.

Biographies

Andreas Berl obtained his Ph.D. at the University of Passau (Germany) in 2011. He is currently working as researcher in the Computer Networks and Communications group at the University of Passau, chaired by Professor Hermann de Meer. His research interests include energy efficiency, virtualization, and peer-to-peer overlays. Currently he is involved in the BMBF project "G-Lab_Ener-G – Improving the Sustainability of G-Lab through Increased Energy Efficiency" and in the EU projects "FIT4Green – Federated IT for a sustainable environmental impact" and "All4Green – Active collaboration in data centre ecosystem to reduce energy consumption and GHG emissions (FP7)". Andreas Berl is member of the EU Network of Excellence "EuroNGI/EuroFGI/EuroNF – Design and Engineering of the Next Generation Internet" and the COST Action IC0804 "Energy Efficiency in Large Scale Distributed Systems". In 2009 he had a DAAD scholarship at Lancaster University, UK, supervised by Professor David Hutchison.

Gergö Lovász received his master degree in computer science in 2008 at the University of Passau (Germany). Currently, he is Ph.D. student at the Chair of Computer Networks and Communications headed by Professor Hermann de Meer at the University of Passau. His main research area is energy efficiency in large-scale distributed systems. Currently he is working on the research project "G-Lab_Ener-G", funded by the German Federal Ministry of Education and Research (BMBF). He is member of the European Network of Excellence EuroNF and the COST Action IC0804 "Energy Efficiency in Large Scale Distributed Systems". In 2010 and 2011 he was local organization chair of the e-Energy conference series on energy-efficient computing and networking. At e-Energy 2012 he was member of the TPC.

Hermann de Meer is currently appointed as Full Professor of computer science (Chair of Computer Networks and Communications) at the University of Passau, Germany. He is director of the Institute of IT Security and Security Law (ISL) at the University of Passau. He had been an Assistant Professor at Hamburg University, Germany, a Visiting Professor at Columbia University in New York City, USA, Visiting Professor at Karlstad University, Sweden, a Reader at University College London, UK, and a research fellow of Deutsche Forschungsgemeinschaft (DFG). He chaired one of the prime events in the area of Quality of Service in the Internet, the 13th international workshop on quality of service (IWQoS 2005, Passau). He has also chaired the first international workshop on self-organizing systems (IWSOS 2006, Passau) and the first international conference on energy-efficient computing and networking (e-Energy 2010, Passau).

Thomas Zettler is Principal System Engineer, responsible for energy-efficiency and power management system-on-chip concepts at Lantiq Deutschland GmbH. He is representative at the European Commission's "European Code of Conduct for Broadband Equipment" and at the U.S. Department of Energy and U.S. Environmental Protection Agency ENERGY STAR Small Network Equipment program. Before this he was Principal in various leading concept and development positions at Infineon Technologies AG and in technology process development at Siemens AG. He holds a Dr. rer. nat. degree and Diploma degree in physics from the University of Hamburg, Germany. He has authored and co-authored numerous publications and is a member of the European Design Automation Association.

Environmental and Economically Sustainable Cellular Networks

Weisi Guo, Siyi Wang and Tim O'Farrell

Department of Electronic and Electrical Engineering, University of Sheffield, Sheffield S10 2TN, UK; e-mail: {w.guo, siyi.wang, t.ofarrell}@sheffield.ac.uk

Received 2 February 2012; Accepted: 10 April 2012

Abstract

The Information and Communications Technology (ICT) infrastructure is recognized as a key enabler to the growth of the global economy. With increased data transfer, there is an unprecedented growth in the associated energy consumption, carbon emissions and operational cost of ICT. In order for operators to increase competitiveness, the challenge is how to satisfy the growing data demand, whilst reducing the energy consumption and costs. This paper considers the wireless cellular network for both outdoor and indoor environments. Novel and theoretical bounds are presented for energy and cost savings.

Investigation results show that with careful redesign of the cellular network architecture, up to 60–70% reduction in energy consumption and 25% in OPEX can be achieved for indoor and outdoor environments. Furthermore, a detailed sensitivity analysis is also presented, which is novel and beneficial to future researchers.

Keywords: energy efficiency, wireless communication, cellular network, cost efficiency.

Journal of Green Engineering, Vol. 2, 273–283.

Figure 1 Energy consumption of (a) ICT and (b) wireless communications as of 2008–2010. A single UK cellular network typically consumes 40 MW. (c) Operational Expenditure (OPEX) of a typical 3G cellular network.

1 Introduction

A recognized key enabler in the modern economy is digital information exchange. The volume of data exchanged in ICT has increased by a factor of 10 over the past 5 years and the associated energy consumption by 20% [1]. One of the most challenging aspects of the ICT infrastructure is the wireless access network, which constitutes 14% of ICT energy consumption (Figure 1a). The global cellular network consists of over 3 million cells and 1.5 billion subscribers. Roughly 70% of the wireless ICT energy consumption is consumed by the outdoor network (Figure 1b), which includes 60TWh of electricity (20 million households). The utility bill is over $10 billion and 40 MT of CO_2 is directly attributed, with a further 500 MT indirectly attributed. Many operators are also pledging to reduce carbon emissions [2].

1.1 Challenges and Solutions

Currently, many operators are considering upgrading their 3rd generation High-Speed-Packet-Access (HSPA) network with the 4th generation Long-Term-Evolution (LTE) network. This is primarily due to the higher spectral efficiency and increased bandwidth of 4G LTE, which amounts to a higher throughput rate [3]. However, it is unclear what the most cost and energy efficient deployment of 4G LTE is. Given an existing LTE reference network deployment [4], the key research questions addressed are:

- *Fundamental Saving Bounds* for energy consumption and operating costs.
- *Outdoor and Indoor* architecture design that achieves the same capacity, but with lower energy consumption and operating costs.

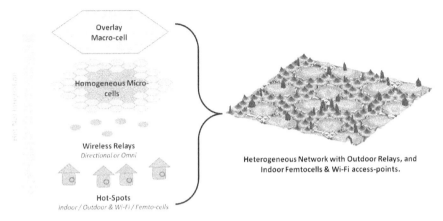

Figure 2 Simulation framework: Heterogeneous network's average received signal-to-interference-plus-noise ratio (SINR).

- *Sensitivity analysis* on each modeling parameter and propose beneficial research directions.

Existing work has primarily focused on specific techniques and their impact on transmission energy efficiency [5–7]. Analysis that considers interference, complete power consumption models and capacity saturation have been lacking [8]. This paper presents fundamental energy and cost saving bounds and demonstrates how savings change with different modeling parameters in order to suggest beneficial areas of research.

2 System Setup

2.1 Monte Carlo Simulation

The simulation results are produced using the MVCE's *VCESIM*, which is a proprietary LTE dynamic system simulator developed at the University of Sheffield for industrial members of the Mobile Virtual Centre of Excellence (MVCE), as demonstrated in Figure 2. The simulator emulates a heterogeneous network of outdoor cells and indoor access-points. Further details of simulation parameters are given in [9].

2.2 Power Consumption

A general power consumption model for a cell is a function of the transmit power (P), load factor (L), radio-head efficiency (μ_{RH}), over-head power

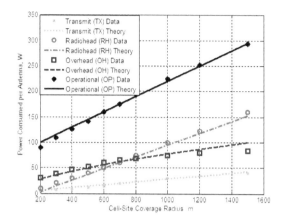

Figure 3 Power consumption variation with cell size with data from [10] and theory from expression (1).

$(P_{\text{cell}}^{\text{OH}})$ and backhaul power (P_{BH}):

$$P_{\text{cell}} = \frac{P}{\mu_{\text{RH}}} L + P_{\text{cell}}^{\text{OH}} + P_{\text{BH}} \approx 0.1 r_{\text{cell}} L + r_{\text{cell}}^{0.62} + 50, \qquad (1)$$

where the radio-head power can be defined as: $P_{\text{cell}}^{\text{RH}} = (P_{\text{max}}/\mu_{\text{RH}})\rho$. The load factor is defined as the ratio between the offered traffic rate and maximum achievable cell capacity: $L = R_{\text{offered}}/R_{\text{cell}}$. Using empirical data from [10], the expression can be shown to be a function of the cell coverage radius (r_{cell}), as shown in Figure 3.

2.3 Operational Cost

As shown in Figure 1c, the electricity bill accounts for up to 16% of the total operational expenditure (OPEX). Global annual electricity bills ($10 billion) make up 45% of the operational and maintenance bills ($22 billion). The OPEX is dominated by the following: site leasing costs $(\wp_{\text{cell,rent}})$, backhaul rental $(\wp_{\text{BH}})$ and electricity bills $(\wp_{\text{bill}})$. The annual OPEX can be written as:

$$\text{OPEX} = N_{\text{cell}}(\wp_{\text{cell,rent}} + N_{\text{BH}}\wp_{\text{BH}} + E_{\text{cell}}\wp_{\text{bill}}), \qquad (2)$$

where N_{BH} is the number of back-hauls per cell, E_{cell} is the energy consumed by the cell [11, 12].

3 Energy and Cost Saving Bounds

3.1 Energy Saving Bound

The paper defines the energy reduction gain (ERG) as

$$\text{ERG} = 1 - \frac{P_{\text{cell,reference}}}{P_{\text{cell,test}}}.$$

Given a *Fixed Deployment* of cells, the paper considers a certain technique that can improve the capacity of each cell by a factor f, which reduces the load of the improved cells by $L = 1/f$. The ERG achieved is:

$$\text{ERG}^+_{\text{RAN,Fixed}} = (\frac{f-1}{f})\Omega, \tag{3}$$

where $\Omega = (\frac{P}{\mu_{\text{RH}}})/P_{\text{cell}}$. The bound approaches 40–60% for $f \rightarrow \infty$. Therefore, one can conclude that for a fixed deployment of cells that have a capacity gain of f, the energy saving gain bound is limited by the ratio between radio-head and total power consumption (Ω).

Given a *Re-Deployment* of higher capacity cells, fewer cells are needed to achieve the same offered load. Assuming that the power consumption change per cell is negligible, the resulting ERG achieved is:

$$\text{ERG}^+_{\text{RAN,Re-Dep.}} \sim \frac{f-1}{f}, \tag{4}$$

which approaches 100% for $f \rightarrow \infty$. Therefore, one can conclude that for a re-deployment of cells, the energy saving is dependent on the capacity gain. The ERG simulation results (symbols) and bounds (lines) are shown in Figure 4a. The simulation results approach their respective bounds when the capacity gain is large $f > 10$.

3.2 Cost Saving Bound

The paper defines the cost reduction gain (CRG) as

$$\text{CRG} = 1 - \frac{\text{OPEX}_{\text{reference}}}{\text{OPEX}_{\text{test}}}.$$

Given a *Fixed Deployment* of cells, the CRG is:

$$\text{CRG}^+_{\text{RAN,Fixed}} = (\frac{f-1}{f})\Gamma, \tag{5}$$

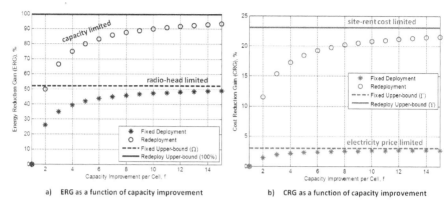

a)   ERG as a function of capacity improvement   b)   CRG as a function of capacity improvement

Figure 4 Energy and cost savings as a function of capacity improvement: symbols indicate simulation results and lines indicate theoretical bounds.

where

$$\Gamma = \frac{\wp_{\text{bill}} \frac{P}{(f\mu_{\text{RH}})}}{\text{OPEX}} t.$$

Over a period of a year ($t = 8760$ hours), the saving is approximately 3% for $f \to \infty$. Therefore, one can conclude that the cost saving gain bound is limited by the ratio between electricity price and the total OPEX: $\wp_{\text{bill}}/\text{OPEX}$.

Given a *Re-Deployment* of higher capacity cells, the paper assumes that each cell needs proportionally more back-hauls. The CRG is:

$$\text{CRG}^+_{\text{RAN,Re-Dep.}} \sim \left(\frac{f-1}{f}\right)\Upsilon, \tag{6}$$

where $\Upsilon = \wp_{\text{cell,rent}}/(\wp_{\text{cell,rent}} + N_{\text{BH}}\wp_{\text{BH}})$. The CRG approaches 26% for $f \to \infty$. Therefore, one can conclude that the cost saving gain bound is limited by the ratio between cell site rental price and the total OPEX. The CRG simulation results and bounds are shown in Figure 4b.

4 Green Cellular Network

4.1 Outdoor Network

The paper analyzes the outdoor and indoor networks separately, but does consider their mutual interference. The overall Heterogeneous RAN (Het-Net) has the following elements (Figure 2): micro-cells, wireless Decode-and-Forward (DF) relays and indoor femto-cell access-points (FAPs). All the

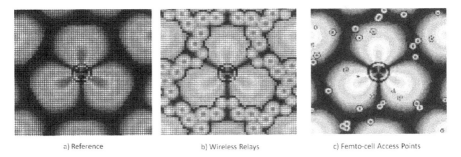

a) Reference   b) Wireless Relays   c) Femto-cell Access Points

Figure 5   Average received SINR for: (a) reference homogeneous, (b) Het-Net with relays, (c) Het-Net with FAPs.

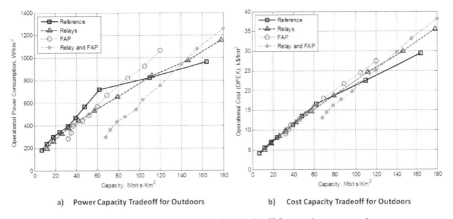

a)   Power Capacity Tradeoff for Outdoors   b)   Cost Capacity Tradeoff for Outdoors

Figure 6   Power, cost and capacity tradeoff for outdoor networks.

elements are co-frequency and mutually interfere with each other. An example of the outdoor average received SINR map is shown in Figure 5. The reference network considered is a homogeneous deployment of micro-cells [3]. The paper introduces heterogeneous elements in the form of 6 relays per cell [12] and 10 FAPs per square km.

The results in Figure 6 show the tradeoff between increasing capacity (through higher cell density) and the associated higher power consumption and OPEX. By introducing heterogeneous network elements, the tradeoff can be both improved and deteriorated. In general, the conclusion is as follows:

- For low-medium traffic loads (30–60 Mbit/s/km^2), a heterogeneous RAN of macro-cells with wireless relays and/or FAPs is more beneficial than a homogeneous deployment.

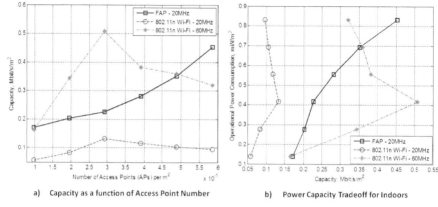

a)   Capacity as a function of Access Point Number   b)   Power Capacity Tradeoff for Indoors

Figure 7 Power and capacity tradeoff for indoor networks.

- For high traffic loads (80–120 Mbit/s/km^2), a homogeneous RAN of pico-cells is more beneficial than a Het-Net.

The rationale behind this is that as the cell-size decreases, the interference caused by relays and FAPs becomes more substantial. Therefore, the green strategy should be to either adopt large cell Het-Nets or small cell Homogeneous Networks. For a target capacity of 70 Mbit/s/km^2, up to 60% operational energy and 25% OPEX can be saved by employing a Het-Net approach. This approaches the theoretical limits for energy cost savings shown previously.

4.2 Indoor Network

In the indoor scenario, the paper considers a large office or commercial area. A number of APs are deployed to achieve a certain capacity density and consume a certain power density. The results in Figure 7a show that for a 20 MHz channel, the FAP offers a superior performance to 802.11n, primarily due to a better adaptive modulation and coding scheme. However, the 802.11n can employ up to 60 MHz of bandwidth in reality, and therefore can always deliver a superior capacity-power-tradeoff, as shown in Figure 7b. As the number of APs increases, increased interference can cause a deterioration in average capacity. This causes the effect that for a given capacity requirement, some tradeoff curves exhibit two power consumption values. Between 60–67% energy can be saved by deploying the right number and type of indoor AP.

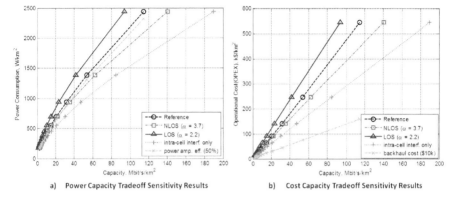

a)   **Power Capacity Tradeoff Sensitivity Results**   b)   **Cost Capacity Tradeoff Sensitivity Results**

Figure 8   Power and capacity tradeoff for indoor networks.

5 Sensitivity Analysis

In the sensitivity analysis, the paper examines the impact of the following:

- Interference: full interference or only intra-cell interference.
- Pathloss Model: pathloss exponent of 2.2 (LOS) or 3.7 (NLOS) [4].
- Power Model: power amplifier efficiency (35 or 50%) [13].
- OPEX Model: backhaul rental cost ($10000 or $40000) [12].

The values are chosen based on existing literature and research on energy efficiency and cellular network modeling. The impact of the parameter values is shown in Figure 8. The results show that in terms of power consumption, the effect of interference modeling has the greatest impact, whereas backhaul cost has the greatest impact on OPEX results. These two areas represent the most promising areas of research for sustainable cellular network design.

6 Conclusions

The paper has demonstrated theoretical and simulation results for the achievable energy and cost savings of a cellular network. In terms of energy saving, the lower-bound is limited by the radiohead consumption (40–60%), and the upper-bound is limited by the capacity gain (up to 100%). In terms of cost saving, the lower-bound is limited by the electricity price (3%), and the upper-bound is limited by the site rental costs (25%).

The simulation results showed that the proposed outdoor and indoor heterogeneous network can reduce energy consumption by 60-70% and OPEX by 25%. This approaches the theoretical bounds and can provide the foundation to a sustainable green architecture. Furthermore, the sensitivity analysis has shown that interference mitigation and backhaul rental costs are the most promising areas of research for energy and cost efficient sustainable cellular networks.

Acknowledgement

The work reported in this paper has formed part of the Green Radio Core 5 Research Programme of the Virtual Centre of Excellence in Mobile and Personal Communications, Mobile VCE. Fully detailed technical reports on this research are available to Industrial Members of the Mobile VCE (www.mobilevce.com)

References

[1] G. Fettweis and E. Zimmermann. ICT Energy Consumption - Trends and Challenges. In *Proceedings of IEEE Wireless Personal Multimedia Communications*, Finland, September 2008.

[2] Sustainability Report 2010-2011. Vodafone Group Plc, Technical Report, 2011.

[3] Ericsson. Summary of Downlink Performance Evaluation. 3GPP TSG RAN R1-072444, Technical Report, May 2007.

[4] 3GPP. TR36.814 V9.0.0: Further Advancements for E-UTRA Physical Layer Aspects (Release 9). 3GPP, Technical Report, March 2010.

[5] C. Xiong, G. Li, S. Zhang, Y. Chen, and S. Xu. Energy- and spectral-efficiency tradeoff in downlink OFDMA networks. *IEEE Transactions on Wireless Communications*, 10(1):3874–3885, November 2011.

[6] S. Tombaz, A. Vastberg, and J. Zander. Energy and cost efficient ultra-high capacity wireless access. In *Proceedings of IEEE Green Net Workshop*, May 2011.

[7] F. Heliot, O. Onireti, and M. Imran. An accurate closed-form approximation of the energy efficiency-spectral efficiency trade-off over the MIMO Rayleigh fading channel. In *Proceedings of 2011 IEEE International Conference on Communications (ICC)*, June 2011.

[8] Y. Chen, S. Zhang, S. Xu, and G. Li. Fundamental trade-offs on green wireless networks. *IEEE Communications Magazine*, 49(6), June 2011.

[9] W. Guo and T. O'Farrell. Green cellular network: Deployment solutions, sensitivity and tradeoffs. In *Proceedings of IEEE Wireless Advanced Conference (Wi-Ad)*, London, UK, June 2011.

[10] G. Auer, V. Giannini, I. Godor, P. Skillermark, M. Olsson, M. Imran, D. Sabella, M. Gonzalez, C. Desset, and O. Blume. Cellular energy efficiency evaluation framework. In *Proceedings of 2011 IEEE Vehicular Technology Conference (VTC Spring)*, pp. 1–6, May 2011.

[11] A. Furuskar, M. Almgren, and K. Johansson. An infrastructure cost evaluation of single- and multi-access networks with heterogeneous traffic density. In *Proceedings of IEEE Vehicular Technology Conference*, May 2005.

[12] E. Lang, S. Redana, and B. Raaf. Business impact of relay deployment for coverage extension in 3GPP LTE-advanced. In *Proceedings of IEEE International Conference on Communications (ICC)*, June 2009.

[13] K. Mimis, K. Morris, and J. McGeehan. A 2GHz GaN class J Power Amplifier for Basestation Applications. In *Proceedings of IEEE Symposium for Radio and Wireless (RWS)*, Phoenix, USA, January 2011.

Biographies

Weisi Guo graduated from the University of Cambridge with BA, MEng, MA and PhD engineering degrees. He is currently a wireless communications research assistant at the University of Sheffield, with research interests in: energy efficiency, cooperative techniques, self-organizing networks, and cellular architectures. He has published over 30 papers and journals and is the author of the VCESIM LTE System Simulator.

Siyi Wang graduated from the University of Leeds and is currently a PhD student at the University of Sheffield. His research interests include: indoor-outdoor network interaction, small cell deployment, and theoretical frameworks for complex networks.

Tim O'Farrell is the Chair in Wireless Communication at the University of Sheffield and the Academic Coordinator of the MVCE Green Radio Project. His research activities encompass resource management and physical layer techniques for wireless communication systems. He has led over 18 major research projects and published over 230 research outputs, including 8 patents.

Ecological Alertness of Cypriot Businesses

S. Louca, K. Matsikaris and I. Stavrides

School of Business, University of Nicosia, 1700 Nicosia, Cyprus; e-mail:
louca.s@unic.ac.cy, k.matsikaris@gmail.com, jake.st4@gmail.com

Received 29 February 2012; Accepted: 1 April 2012

Abstract

Participation in environmentally friendly or green initiatives for ensuring that all processes, products, and manufacturing activities adequately address current environmental apprehensions has become mandatory in today's global operating environment by any organization. The ability of a business to meet the needs of the present world without compromising the sustainability and the well being of the future generations to meet their own needs is a social and ethical challenge. A lot of debate is being done over the last decades about the impact of consuming and wasting products on the environment. With the constant increase of consumers' awareness on the matter, businesses and governments are asked to provide solutions to these questions. This growing trend concerning "green businesses" forces companies to develop and implement strategies to minimize environmental impact that their products and services have. The focus of this paper is to investigate the responsiveness of Cypriot businesses in the global call for ecologically friendly initiatives along with their impact on their performance, on their business strategy and on their consumers. The findings show that companies in Cyprus are pursuing various environmental initiatives involving mainly their recycling habits and issues, certification and energy preservation.

Keywords: business sustainability, ecological alertness, green initiatives, Cyprus.

Journal of Green Engineering, Vol. 2, 285–303.

1 Introduction

> We shall require a substantially new manner of thinking if mankind
> is to survive.
> Albert Einstein.

It is a fact that the environment is in severe danger, mostly from human intervention and misuse of its natural resources. There is a fixed amount of resources, many of which have already been extinct. The planet's bio-capacity of regenerating its natural ecosystems is falling, thus limiting its ability to sustain life itself. On top of that, the rapid growth of the world's population, scarce the already limited resources. According to recent demography, world population grows at a geometric rate the past two thousand years. In 2000 it reached 6 billion and it is estimated that it will reach 10 billion by the year 2080. Ecosystems like forests, oceans rangelands are threatened, while others are nearly collapsed. According to World Centric, a company with sole mission to educate the public and prevent the environmental misuse, forestlands are cut down at a rate of 375 km^2 each day and 80% of the planets initial forestlands have been destroyed, while 75% of our planets fish stock is either exploited or recovering [16].

Still all is not lost. As human beings, our goal is to preserve and maintain our legacy for our children to enjoy. If we are to pass on to our future generations the same standards and quality of life that we have enjoyed, we must change the way we conceive our environment and the usage of its natural resources. One of the many growing trends in public concern, as well as businesses practices about the environment is "going green", making a healthy impact on all living things around us; adopting in other words, a more environmentally friendlier approach. Going green can have various applications to individuals and businesses; from reducing energy use by supporting and using alternative energy sources (solar and wind energy), reducing greenhouse effect by reducing the emission of gas in to the atmosphere, to recycling paper and tin cans in your house. Due to the fact that this trend is gaining more and more public awareness, thus increasing the demand for environmentally friendly products, for individuals and businesses alike, governments are realizing that the need to design and implement strategies to reduce the disastrous impact that business activities have on the environment is dire.

To this end, initiatives have started to be taken; although the technological improvement over the last decades has improved our quality of life, it had negative effect on the environment, like global warming and greenhouse effect. In order to cope with that, jobs intended to aid the environment have been created, the so called green collar jobs, including hazardous material clean-up, development of non-toxic cleaning products and others [17]. Also businesses began adopting environmentally friendly initiatives; by converting all of their records in electronic form and promoting paperless communication through the usage of Internet and Intranets, reducing the waste office and paper supplies intensely. Also by selling their old electronic equipment instead of disposing it, saves the companies a great deal of money and contributes to the preservation of the environment.

The simplest and easiest way to contribute to this cause is by recycling. According to the "Go Green Initiative", a company that provides to individuals and businesses the training and means to develop an environmentally friendly attitude, the greatest threat that the natural ecosystem faces, derives from the things we throw away in our everyday consumption. For example batteries and electronic devices contain toxic fluids that when spilled and absorbed by the soil, it can pollute everything from the soil we grow our food, to the underground water streams which provide our household water supply [18].

The benefits from recycling are tremendous; the numbers speak for themselves. According to Green Waste, a recycling company, each ton of paper recycled can save among others 7,000 gallons of water, 380 gallons of oil and eliminate 60 pounds of air pollutants [19]. Materials like glass and plastic can be completely recycled again and again, yet tons of them end up in landfills, where the average time to decompose and absorbed by the environment is nearly 700 years. The most alarming fact though is that if someone was to check a typical dustbin, he would discover that 60 to 70% of the waste can be recycled, yet they end up in landfills! Aluminum is one other material that recycling it saves energy (since it takes 100 times more energy to create it from raw metals than to recycle it) and time. Characteristically, the energy saved from recycling aluminum, can light a small city for nearly five years.

In Cyprus businesses have started to take steps in adopting their own environmentally friendly attitude. This survey tends to identify the areas and the magnitude of these initiatives and at the same time point out areas that can be further analyzed and improved. The first part of the paper is based on existing literature on the topic of green and sustainable operations of various companies in the world. The second part relates to the roadmap and ecological alertness of Cypriot businesses based on an analysis of primary data collected from 141 companies in Cyprus.

2 Ecological Alertness

Over the last decade the misuse of the environment has reached dangerous levels, intensifying public concern over the matter and making the need for a "greener" way of doing business imperative. Companies, especially midsized ones turn to green IT and sustainable growth strategies, not only for the sake of the planet but also as a mean to gain competitive advantage. The new "green" advances in technology provide the companies a smarter and more efficient way to compete, resulting also in reduction of energy costs and organizational complexity. Customers and other stakeholders exert pressures on organizations to improve their ecological attitude. Companies recognize that if they fail to deliver on this, it will transform into a negative impact on profit. Furthermore, they are discovering that green initiatives offer costs savings benefits while restructuring the organization, meeting stakeholder demands and complying with laws and regulations. Many governments are introducing antagonistic ecofriendly policies, encompassing everything from greenhouse gas reduction and natural resource protection to clean power initiatives and incentives for energy efficiency [22].

Although much effort has been proved successful in terms of conceptualizing and understanding the environmentally friendlier way of conducting businesses, it continues to be a complex matter. Many environmental factors such as quality requirements, cost feasibility and customer satisfaction needs to be addressed. The progress made on the subject can be identified by the creation and evolution of environmental management; from pollution control and risk management in the 1970s, to pollution prevention in the 1980s to the launch of the ISO 14000 series and the emergence of industrial ecology approaches [4]. This constant evolution has led to the belief that environmental management is a key area for companies that want to be competitive in the modern global economy [5]. Nevertheless, many organizations all over the world companies are still facing the challenge of strict sustainability demands pursuing environmental leadership. Most environmental initiatives are uncertain, and there are also many other non-environmental factors issues that need to be addressed such as required investment, customer satisfaction and quality requirements [10].

Organizations are investing in green IT initiatives, not only to improve on their efficiency, but also to become more responsible organizations towards the environment and the society. Many software organizations have created a green information technology framework to reduce their contribution to the greenhouse effect, their 'carbon footprint' in other words. Carbon footprint is the "measure of the amount of greenhouse gases, measured in units of carbon dioxide, produced by human activities" [7]. This turn towards the reduction of the carbon footprint is due to the regulations formulated mostly by the European Union in an effort to help companies become more environmentally friendly. Regulations like Restriction of Hazardous Substances (RoHS) and Waste Electrical and Electronic Equipment (WEEE) "address the increasing electronic and electrical waste by restricting the use of toxic substances and flame-retardants by the manufacturers of electrical and electronic equipment to a particular level. This regulation also addresses the recycling programs for the manufacturer's products" [1, 2, 19].

Large organizations are taking it upon themselves to go green in this day and age. They realize that the impact that such decisions will enhance office morale by modernizing company practices, bringing the organizations into the 21st century like just about nothing else can. For example, IBM is creating strategies to provide technologies for smarter planet. In order to achieve this, it designs their IT framework based on green initiatives, hence the representative "Green IT". Green IT initiatives include [22]:

- *Virtualization & Consolidation*: Initiatives in this area include server virtualization and consolidation, storage consolidation and desktop virtualization for improving cost and energy efficiency through optimized use of existing and new computing and storage capacity, electricity, cooling, ventilation and real estate.
- *Energy Efficiency*: Initiatives in this area include improvements in the server room, IT energy measurement, printer consolidation, and PC power management. Such actions aim at energy efficiency and thus cost savings.
- *Travel Reduction*: Initiatives in this area include remote conferencing & collaboration and telecommuting for reducing travel costs.
- *Asset Disposal:* Recycling of e-waste. These wastes are most hazardous to people and the environment, since left exposed to sun and water, they release toxic gas.

Furthermore, companies are investing into the usage of alternative or renewable energy to address and reduce its operating energy consumption. Solar, wind, tidal, geothermal and biofuel energy are some examples. Solar energy deriving from sunlight is most commonly used, since it is the easiest of them all to harness, with solar energy gaining much popularity. They are operating Green Data Centers where, energy consumption is reduced through various initiatives, like virtualization, cloud and grid computing.

Virtualization is the process of abstracting various computing resources and consolidating them to a single physical network [14]. That is, it is the process of managing all the services and information in one virtual network, resulting in minimizing the natural resources used as well as the greenhouse gas emissions. Grid computing combines "computer resources from multiple administrative domains to reach common goal" [15].

Other software companies like Microsoft and Cisco, as well as the CERN research center are promoting this system, since the reduction of energy consumption during data processing is substantial. Another aspect where businesses try to implement green initiatives and improve their environmental performance is within the entire supply chain management, the so called green supply chain (GSC) initiatives. Supply chain management is the process of introducing raw materials into a business, converted in to final products and delivered to the end-consumer [4], with the involvement of the extraction and exploitation of the natural resources [8]. The addition of the green concept in the supply chain was introduced in the 1990s when manufacturers where pressured to address the Environmental Management in their operations. According to Porter and van der Linde [11] the basic reasons for implementing GSC can be resource saving, waste elimination and productivity improvement. Furthermore, three approaches in GSC where suggested; reactive, proactive and value seeking [12, 13]. In the reactive approach, organizations devote only minimum resources to environmental management and try to lower their environmental impact by implementing "end of pipeline" initiatives. In the proactive approach, organizations start to enact environmental laws and also commit resources in recycling and green design. In the value-seeking approach organizations incorporate activities like ISO implementation and green purchasing in their business strategy as initiatives.

Every year, Newsweek ranks the 500 largest US companies and the 500 largest global companies in terms of their green scores. Company size is evaluated according to revenue, market capitalization, and number of employees. The green score is derived from three component scores: Environmental Impact Score (EIS), Green Policies Score (GPS), and the Reputation Survey Score (RSS), weighted at 45, 45, and 10%, respectively. The companies' green scores are ranked on a scale from 100 (highest performing) to 1 (lowest performing) [3,23]. The rankings for the top 15 for 2011 are shown in Table 1.

Table 1 Top 15 green companies globally for 2011 [23].

Rank	Company	Country	Industry Sector	Green Score	Environ- mental Impact
1	Munich Re	Germany	Financials	83.6	87
2	IBM	United States	Information Technology & Services	82.5	78.8
3	National Australia Bank	Australia	Financials	82.2	80.6
4	Bradesco	Brazil	Financials	82.2	88.1
5	ANZ Banking Group	Australia	Financials	80.9	84.9
6	BT Group	United Kingdom	Telecommunications	80.4	76.2
7	Tata Consultancy Services	India	Information Technology & Services	79.1	73.3
8	Infosys	India	Information Technology & Services	77.3	75.3
9	Philips	The Netherlands	Capital Goods	77.2	59.7
10	Swisscom	Switzerland	Telecommunications	77	77
11	Societe Generale	France	Financials	76.6	74.4
12	Bell Canada Enterprises	Canada	Telecommunications	76.5	73.5
13	Fujitsu	Japan	Technology Equipment	76.1	67.5
14	Wal-Mart de Mexico	Mexico	Retailers	75.9	63.6
15	Hewlett-Packard	United States	Technology Equipment	75.8	66.7

3 Waste Management in Cyprus

Cyprus is the third largest island in the Mediterranean after Sicily and Sardinia with an area of 9,251 km^2, situated at the north-eastern end of the East Mediterranean basin. Its estimated population (estimated September 2009) is 796.8 thousands. The area of the Republic of Cyprus under government control has a market economy dominated by the service sector, which accounts for 78% of GDP. The Republic of Cyprus has *de juris sovereignity* over the whole island but de facto has no control over the northern part, as it is occupied by Turkey. As a result, the challenges faced by governmental and private organizations involved in ecological initiatives are even greater.

As a result of globalization and the accession of Cyprus to the European Union, on May 1, 2004, new conditions were created which dictate appropriate orientation of development of the Cyprus businesses. The Cyprus legislation on waste management and special management of packaging waste was prepared based on relevant European legislation and directives and implemented in 2002. It is obligatory, by law, that every company which puts on the market packaged goods to collect back the packaging and recycle it. It was not, however, put in force till late 2009 with the formation of Green Dot. According to the EU Directives enforceable laws, regulations and directives to motivate organizations follow them have been placed. Thus, Green Dot was formed. Green Dot is a non-profit organization which undertakes to collect the packages from manufacturers. This is allowed by law and is called "collective system". The charge is based on packaging material and the quantity put on the market [2].

The majority of the industry in Cyprus follows the Green Dot collective system as it is the only Collective legal entity in Cyprus. It offers Companies, based on a fee, to settle their recovery and recycling obligations. The Green Dot Cyprus is a member of the big global family of collective management of packaging, the Packaging Recovery Organization, Europe (PRO-EUROPE) [21], which now houses under the umbrella of the 26 relevant organizations from Europe and America. Companies that are not a member of Green Dot may follow an independent system, still being subject to the same regulations.

Cyprus Strategic Plan for the management of waste led to the establishment of four regional centers for integrated management of solid waste as well as the restoration of existing uncontrolled landfills. The first regional centers, in Larnaca and Famagusta Districts became operational in 2010. Based on the report prepared by the Ministry of Agriculture, Natural Resources and Environment [1] for 2008 and delivered in August of 2010, the categories concerned are paper, plastic, metals, glass, wood, mixed and other with the total number of tones being collected reaching approximately 87,466. The "other" category reaches 4606 tones and we assumed that part of it is the e-waste collected for 2008.

4 Methodology

For data gathering, a quantitative methodology was employed based on a survey with 141 participating organizations within Cyprus. The target organizations came from the public, semi-public and private sectors on the island. The initial survey was carried out in December 2011.

The quantitative methodology is defined as the systematic investigation of social phenomena via statistical, mathematical and computational techniques in order to develop and employ mathematical models and hypotheses related to the phenomena at hand. The results of such a process, the data are in numerical form. In other words, during the quantitative research, questions asked are narrow and specific and answers are retrieved in mathematical form and with the help of statistics the researcher hopes to derive results that can be applied to a larger population. Qualitative research on the other hand collects word answers through broad questions, in order to identify patterns for the specific participants.

The quantitative method was preferred, since the intention of this research was to establish a statistical relationship among the Cyprus businesses concerning ecological friendly initiatives. The project framework was based on data collected from 141 small and middle sized businesses from various sectors in Cyprus through an online questionnaire used in the form of a survey. The content of the questionnaire was created in such a way in order to identify the measure of awareness that businesses in Cyprus have of the impact that their activities have on the environment and how this concern is converted into action. It constituted of questions concerning company demographics, recycling habits and issues, certification, energy preservation, and environmental impact and policies.

The data were then analyzed with the help of SPSS in order to identify relationship between the data in the form of tables and graphs.

5 Results

In this section, the results of the research are presented giving emphasis on the opportunities for green initiatives and the responsiveness of the Cyprus businesses.

5.1 The Sample Structure

The number of organizations participating in the survey amounted in 141. As shown in Table 2, 12.8% are operating in the public sector, 27% in the semi-public sector and 60.2% in the private sector. Table 2 shows the specific business sectors in which the 141 companies are operating.

As shown in Figure 1, the companies surveyed are located in the major cities of Cyprus. 42.6% are located in the capital of the island, Nicosia, 19.1% in Larnaca, 17.7% in Limassol, 16.3% in Paphos and 4.3% in the Famagusta area.

Table 2 Sample structure.

	Percentage	Public	Semi-public	Private
			Sector	
Agriculture, forestry and fishing	5.7	1.4	1.4	1.8
Arts, sports and recreation	5.0	0.7	1.4	2.8
Catering and accommodation	7.1		0.7	6.4
Construction	7.1	0.7	3.5	2.8
Education	6.4	2.1	0.7	3.5
Health and social care services	4.3	0.7	2.1	1.4
IT and telecommunications services	12.1	2.1	2.1	7.8
Manufacturing	9.9	0.0	3.5	6.4
Media and creative services	9.9	0.7	2.8	6.4
Mining, energy and utilities	6.4	1.4	2.1	2.8
Personal services	10.6	0.0	0.0	10.6
Professional and business services	7.8	2.1	3.5	2.1
Retail, hire and repair	7.8	0.7	2.8	4.3
Total	100.0	12.8	27.0	60.2

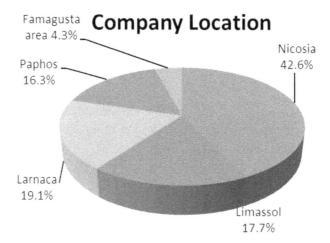

Figure 1 Company location.

5.2 Recycling Habits

Traditionally waste management focused on hauling wastes out of the city boundaries and dumping them in some remote area. This is in conformity with the "out of sight out of mind" philosophy, dated back to the 13th century [25]. However, with the ever increasing capacity of waste due to the expansion of urban centers, which implies not only increased collection, transportation and disposal costs, but also a negative impact on the en-

Table 3 Amount of recycling containers used and type of containers used.

Liter	Amount (in %)	Type of container (in %)
1100	1.4	7.8
660	2.8	19.1
360	11.3	24.8
240	42.6	24.1
Sacks	41.8	24.1
Total	100.0	100.0

Table 4 Frequency of collection of recycled waste and other waste.

Period	Recycled Waste (in %)	Other Waste (in %)
Weekly	52.2	98.6
Bi-weekly	34.1	1.4
Monthly	13.7	13.5

vironment. Recycling has become a sustainable approach to solid waste management and to an environmentally friendly attitude.

Our research regarding the recycling habits of the Cyprus businesses has shown that they have already started to adapt their behaviors and habits in order to be more environmentally and ecologically conscious with their actions. The majority of these businesses stated that their recycling waste does not exceed the 240 liters per collecting period. The 240-liter containers are considered to be the minimum in size of recycling waste. According to our survey, the other waste produced is mainly collected in 360–660 liters containers (see Table 3). The main reason for this big difference between recycle waste and other waste can be attributed to the frequency of collection of the waste by the proper authorities. Epigrammatically, we can mention that 52.2% of the companies that participated answered that recycled waste is collected weekly and 34.1% every two weeks, while on the other hand, 98.6% of the companies answered that other waste is collected weekly (see Table 4).

According to Figure 2, the majority of the recycling waste corresponds to office paper with 63%, tetra packs with 66%, plastic with 87.2% and tin cans with 39%. E-waste which is also considered to be harmful for the environment is only at 27.7%. From the data of this table we see that companies focus mostly on recycling paper and plastic. Although the volume of the paper and plastic recycled is high enough, it would be wise if companies could include in their recycling programs other materials widely used in offices,

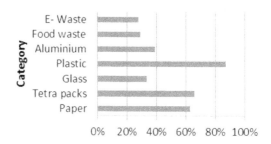

Percentage

Figure 2 Recycled products.

Table 5 Green certification.

Certificate	Percentage
Elemental Chlorine Free Certification	3.5
Green Standard Certification Program	2.8
Lead Free Certification Scheme	0.7
GSCP recognizes Green Living Ideas	5.0
United Academy of Businesses, UK	3.5
Quality Leader Forum	2.8
ISO14000 Series Environmental Management Systems	71.6

like electronic waste, which could also save companies additional expenses on purchasing new electronics.

As shown in Figure 3, the majority of the businesses that participated in our survey have also stated that the main reason for not recycling more is the inefficiency of the collection services to respond to the companies' recycling volume. Also many companies stated that they do not have the efficient time or budget to recycle more.

5.3 Certification

As shown in Table 5, the most common Green Initiative Certificate with 71% among the companies is the ISO 14000 Series Environmental Management Series. 90%, however, stated they do own some kind of green certificate, but only 27% use this information in their marketing campaign. Table 6 presents the barriers for green certification, the main reason being the acquisition costs.

Figure 3 Reasons for not recycling.

Table 6 Reasons for not owning a green certificate.

Reason	Percentage
Cost of acquisition	10.6
No knowledge of such certifications	2.1
No real benefit	7.1
Not required by legislation	4.3
Other	4.3

5.4 Energy Preservation and Efficiency

As far as proper heating and cooling insulation installed in the business facilities, 35.5% of the companies reserve 5–10% of energy, while 18.4% have no energy saving insulation installed in their buildings (see Figure 4).

The most common product that derives from recycling waste and companies use are mostly office supplies, like toilet and office paper, refillable pens, one-use plastic cups and others. Also environmental friendly products, like non-toxic cleaning supplies are commonly used in companies, while the percentage of companies that do not use any kind of energy efficient products are close to 20%. The table that follows shows in detail the percentage of each energy efficient product is used.

The main barriers that prevent companies from using energy efficient products are the cost of acquiring them and the lack of quality. The reason for the high cost of those products is that since in Cyprus there are no recycling factories to process recycling materials and produce energy efficient products, the cost of importing these products is high and most of the companies cannot afford these expenses and resolve to non-recycled products (see Table 8).

Although that 95% of the companies are concerned with the impact of their business activities on the environment, 54.6% do not have an efficient

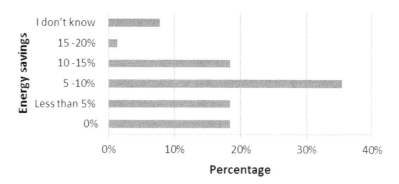

Figure 4 Energy savings due to building insulation.

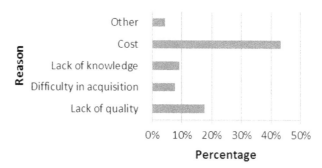

Figure 5 Reasons for not using efficient products.

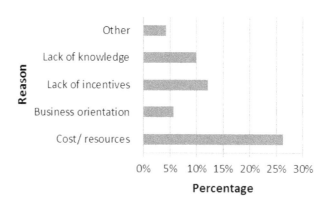

Figure 6 Barriers to adopting efficient environmental policies.

Table 7 Energy efficient products used by companies.

Type	Percentage
Office supplies	54.6
Alternative energy sources	37.6
Vehicle fuels	16.3
Environmental friendly products	39.7
None of the above	20.6
Other	2.1

Table 8 Reasons for not using re-cycled products.

Reason	Percentage
Lack of quality	17.7
Difficulty in acquisition	7.8
Lack of knowledge	9.2
Cost	43.3
Other	4.3

environmental policy. In other words, as humans everyone is aware of the grave impact that our activities have on the environment, but as businessmen we prioritize the profit of our company by any means necessary. That is why the main reason that companies do not adopt an efficient environmental policy is the cost and lack of resources and as result, 79% of the companies are not involved in any kind of community green initiative while the rest 21% did not answer this question.

6 Discussion

The above results show that businesses in Cyprus are concerned with the impact their business operations have on the environment and each of these companies in their own way tries to reduce their waste by recycling. Organizations are attempting to reform their behaviors and habits in order to become more environmentally conscious with their actions, mainly through recycling. Although they are trying to convert their environmental concern to action, they face a number of difficulties towards achieving their goal, mostly due to the inability of the waste recycling companies to meet their recycling volumes, and the investment required for green initiatives.

The first accredited step in the business environment is acquiring a Green Certificate that their products and services are environmentally friendly. Although the majority of the companies own some kind of Green Certificate, only few of them use it as a marketing tool. Since public concern about the environment over the last few years has increased and people in Cyprus have started recycling in their homes, companies should consider using their certificates in their marketing campaigns in order to attract more environmental concerned customers. As far as energy efficiency in terms of cooling and heating business buildings, Cyprus companies do not have the efficient infrastructure to support such installations. To encourage improvements in this sector, it is recommended that that the public authorities should impose environmentally friendly regulations to any new buildings and facilities built.

These facts strengthen the need for more decisive actions. Companies that are in charge of collecting recycled waste should collect them more often, especially from big companies with high volume. In this way, by increasing the volume of recycled waste, it would be financially wise to develop factories that can collect recycled waste and provide companies with energy efficient products, thus reducing the costs of exporting and importing recycled products. In addition, external pressures, such as European regulations on e-waste and on the use of certain hazardous substances, have been driving firms and governments to initiate various ecological actions. It is important that such initiatives must compete against other prospective projects in an organization (marketing campaigns, production capacity expansion, quality improvement, etc.). On the other hand, public pressure, legislation, possible cost reduction and improved company image may be sufficient motivation to improve environmental performance [6]. In addition, economic support (i.e. in the form of tax waives) could further promote ecological friendly actions. Public awareness of environmental protection is improving continuously. The Government and Green Dot have released campaigns to promote this problem to people.

The findings suggest, despite the potential profitability and the existing legislative requirements, that there are no integrated decision frameworks to advise decision-makers about the economic viability of green initiatives.

7 Conclusion

Organizations worldwide apprehend that addressing environmental issues and concerns can influence their well-being and their profits. They are discovering that green initiatives offer costs savings benefits while restructuring the organization, meeting stakeholder demands and complying with laws and regulations. The adoption of green initiatives is an opportunity for an affirmative effect on the environment, but at the same time use resources more efficiently. Furthermore, many governments are introducing antagonistic ecofriendly policies, encompassing everything from greenhouse gas reduction and natural resource protection to clean power initiatives and incentives for energy efficiency. This paper examines the responsiveness of Cypriot businesses in the global call for ecologically friendly initiatives along with their impact on their performance, and on their business strategy. The findings show that companies in Cyprus are pursuing various environmental initiatives involving mainly their recycling habits and issues, certification and energy preservation. However, systemic changes are required to have a larger ripple effect throughout all of society in order to place Cyprus to the global green initiative map. Future work will focus specifically on Green IT initiatives for transforming organizations while complying with laws and regulations.

References

[1] Department of Environment, Ministry of Agriculture, Natural Resources and Environment. Appropriate Description of the data used according to the Article 1 of Commission Decision 2005/270/EC on packaging waste. Delivery date 9 August 2010.

[2] ECO-Logic. A Report on the Implementation of Directive 2002/96/EC on Waste Electrical and Electronic Equipment (WEEE). Institute for European Environmental Policy, April 2009.

[3] P. Bansal and K. Roth. Why companies go green: A model of ecological responsiveness. *The Academy of Management Journal*, 43(4):717–736, 2000.

[4] B. Beamon. Designing the green supply chain. *Logistics Information Management*, 12(4):332-342, 1999.

[5] C.J. Corbett and R.D. Klassen. Extending the horizons: Environmental excellence as key to improving operations. *Manufacturing and Service Operations Management*, 8(1):5–22, 2006.

[6] J. Hall. Environmental supply chain dynamics. *Journal of Cleaner Production*, 8(6):455-471, 2000.

[7] M.L. Walser. Reducing carbon footprint. *International Journal of Enterprise Computing and Business Systems*, 1(2), 2011.

[8] S. Srivastava. Green supply-chain management: A state-of-the-art literature review. *International Journal of Management Reviews*, 9(1):53-80, 2007.

[9] Jamal Fortes. Green supply chain management: A literature review. *Otago Management Graduate Review*, 7, 2009.

[10] B. Nunes and D. Bennett. Green operations initiatives in the automotive industry: An environmental reports analysis and benchmarking study. *Benchmarking: An International Journal*, 17(3):396–420, 2010.

[11] M.E. Porter and C. van der Linde. Green and competitive. *Harvard Business Review*, 73:120–134, 1995.

[12] R.J. Kopicki, L. Legg, L.M.J. Berg, V. Dasappa, and C. Maggioni (Eds.). *Reuse and Recycling: Reverse Logistics Opportunities*. Council of Logistics Management, Oak Brook, IL, 1993.

[13] R.I. van Hoek. From reversed logistics to green supply chain. *Supply Chain Management*, 129–135, 1999.
[14] http://searchnetworking.techtarget.com/generic/0,295582,sid7_gci1310084,00.html.
[15] V. Berstis. Fundamentals of grid computing. From http://www.redbooks.ibm.com/abstracts/sg246778.html?Open. Accessed February 2012.
[16] World Centric, http://www.worldcentric.org/conscious-living/environmental-destruction. Accessed January 2012.
[17] The Green Initiative, http://www.chinatownconnection.com/green-collar-jobs.htm. Accessed January 2012.
[18] Go Green Initiative, http://www.gogreeninitiative.org/. Accessed January 2012.
[19] Green Waste http://www.greenwaste.com/recycling-stats). Accessed January 2012.
[20] http://www.phrases.org.uk/meanings/274400.html. Accessed February 2012.
[21] www.pro-e.org. Accessed November 2010.
[22] http://www-03.ibm.com/press/attachments/GreenIT-final-Mar.4.pdf. Accessed February 2012.
[23] http://www.thedailybeast.com/newsweek/features/green-rankings/2011/international.html.

Biographies

Soulla Louca received her Ph.D. in Computer Science in 1994 from the Illinois Institute of Technology in Chicago. Prior to that, she received a B.A in Computer Science and Mathematics and a M.Sc. in Computer Science from Kalamazoo College and Western Michigan University respectively. She has participated and coordinated numerous projects including National Science Foundation (NSF-USA), Research Promotion Foundation (RPF-Cyprus) and European; has served as a reviewer for various international conferences/journals as well as an ICT expert for the European Commission. Her research interests include socio-economic aspects of Green ICT, e-learning, social integration and digital divide, and e-business Since June 2008, she has been the Chair of the Domain Committee for Information Communication Technologies for the European program COST (Co-operation for Science and Technology, www.cost.eu). Dr. Louca is an Associate Professor at the University of Nicosia in the Department of Management and Management Information Systems.

Kyriakos Matsikaris is a senior student at the University of Nicosia pursuing a Bachelor degree in Management Information Systems. Upon graduation, he plans to follow a career in Management and Human Resources.

Iacovos Stavrides is a senior student at the University of Nicosia pursuing a Bachelor degree in Management Information Systems. Upon graduation, he plans to continue his studies for a Master's degree.

Online Manuscript Submission

The link for submission is: www.riverpublishers.com/journal

Authors and reviewers can easily set up an account and log in to submit or review papers.

Submission formats for manuscripts: LaTeX, Word, WordPerfect, RTF, TXT.
Submission formats for figures: EPS, TIFF, GIF, JPEG, PPT and Postscript.

LaTeX

For submission in LaTeX, River Publishers has developed a River stylefile, which can be downloaded from http://riverpublishers.com/river_publishers/authors.php

Guidelines for Manuscripts

Please use the Authors' Guidelines for the preparation of manuscripts, which can be downloaded from http://riverpublishers.com/river_publishers/authors.php

In case of difficulties while submitting or other inquiries, please get in touch with us by clicking CONTACT on the journal's site or sending an e-mail to: info@riverpublishers.com

www.ingramcontent.com/pod-product-compliance
Lightning Source LLC
LaVergne TN
LVHW012332060326
832902LV00011B/1843